Places of the World

A Study Guide for Achieving Fundamental Geographic Literacy

Sixth Edition

For

Hobbs'

World Regional Geography

Sixth Edition

James W. Lett, Ph. D
Indian River College

BROOKS/COLE
CENGAGE Learning™

Australia • Brazil • Japan • Korea • Mexico • Singapore • Spain • United Kingdom • United States

BROOKS/COLE
CENGAGE Learning™

For product information and technology assistance, contact us at **Cengage Learning Customer & Sales Support, 1-800-354-9706**

For permission to use material from this text or product, submit all requests online at **www.cengage.com/permissions** Further permissions questions can be emailed to **permissionrequest@cengage.com**

Cover Image: © Keren Su/Getty Images

ISBN-13: 978-0-495-39101-2
ISBN-10: 0-495-39101-8

Brooks/Cole
10 Davis Drive
Belmont, CA 94002-3098
USA

Cengage Learning is a leading provider of customized learning solutions with office locations around the globe, including Singapore, the United Kingdom, Australia, Mexico, Brazil, and Japan. Locate your local office at: **international.cengage.com/region**

Cengage Learning products are represented in Canada by Nelson Education, Ltd.

For your course and learning solutions, visit **academic.cengage.com**

Purchase any of our products at your local college store or at our preferred online store **www.ichapters.com**

Printed in the United States of America
1 2 3 4 5 6 7 11 10 09 08

For Teresa,
who was a world of help in writing this book,
and who means the world to me

CONTENTS

Visual Index of World Regions

Planet Earth

Europe

Russia & the Near Abroad

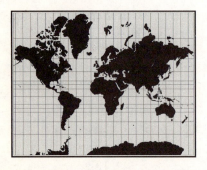

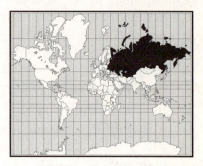

Middle East & North Africa

Monsoon Asia

Oceania

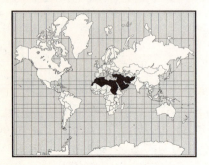
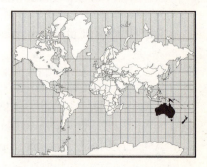

Sub-Saharan Africa

Latin America

United States & Canada

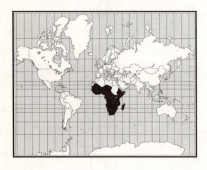
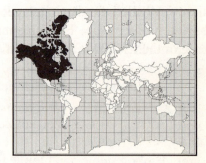

PREFACE
Fundamental Geographic Literacy

In Act II, Scene VII of *As You Like It*, William Shakespeare remarks that "All the world's a stage, And all the men and women merely players." If Shakespeare had been a geographer, he might have observed that the earth is the stage upon which the drama of human existence is played out, and that all of us who are actors in the play need to be familiar with the layout of the theater if we hope to make sense of the stage directions. Making us familiar with the layout of our global theater is a central task of geography: the literal meaning of the word **geography** is "description of the earth."

Geographers rightly object, however, to the perception that their field is concerned with *mere* description, and they frequently lament the widespread association of their discipline with the rote memorization of countries and capitals. As geographers insist, the primary goal of their discipline is not to catalogue the world's places but to identify and explain the connections and linkages between those places. The field of geography is intimately tied to the concept of spatial analysis: if a phenomenon can be mapped, then it's amenable to geographic research and insight. The National Geographic Society has identified six essential elements that define the contemporary discipline of geography (*The World in Spatial Terms*, *Places and Regions*, *Physical Systems*, *Human Systems*, *Environment and Society*, and *The Uses of Geography*), and that's why geographers adamantly maintain that their field involves far more than simply memorizing the names of various locations around the world.

It's certainly true that being able to name and locate the world's principal geographic features won't make you a geographer, any more than being able to name the principal components of an automobile engine will make you a mechanic, or being able to distinguish between a metaphor and a simile will make you a literary critic. However, every competent mechanic needs to know what a transmission is and where it's located, and every capable literary critic needs to be able to recognize various figures of speech. In exactly the same way, every geographer needs to have an underlying knowledge of fundamental place-name geography, which entails being able to name and locate the world's principal geographic features (including the world's major landforms and bodies of water as well as the world's countries). That kind of basic background knowledge might be called fundamental geographic literacy, and while it isn't *sufficient* for geographic analysis, it is absolutely *necessary*. Fundamental geographic literacy is *fundamental* to the field of geography in the strict sense of the term: it is the foundation upon which all geographic understanding is built.

It's not only geographers, however, who need to know where places are in the world and what they're called—*everyone* needs that kind of knowledge. Without a basic knowledge of world geography, it would be impossible to fully understand world history, global economics, international affairs—or virtually anything else that involves or affects human beings. In order to understand news reports about international affairs, for example, it is essential to have a basic framework of geographic knowledge. Developments in the Middle East are the frequent topic of news reports in the United States, but how many Americans know that the terms "Arab" and "Muslim" are *not* synonymous? How many are aware that all Arab countries are Muslim, but that not all Muslim countries are Arab? How many know that Iraq is Arab but Iran is not, or that Turkey and

Afghanistan are Islamic but not Arab? For someone who lacks this kind of fundamental knowledge, news reports from the Middle East must be very confusing indeed.

Regrettably, geographic illiteracy is widespread in the United States, as annual surveys by the National Geographic Society have repeatedly demonstrated in recent decades. In comparison with the citizens of other industrialized countries, Americans rank embarrassingly low in terms of their knowledge of world geography. That fact is widely recognized, but the dangers of geographic illiteracy may not be sufficiently appreciated. There has always been a faction in American society that favors isolationism, and the present is no exception. Contemporary isolationists, like their predecessors, are often motivated by parochialism and xenophobia, and in their disdain for other countries beyond their own borders they sometimes even take pride in their ignorance of world geography. Such pride is misplaced; willful ignorance is almost always detrimental, and certainly so when the subject is the world in which we live.

There are at least three reasons why everyone should strive to achieve fundamental geographic literacy. First, a basic knowledge of world geography provides the necessary framework for learning about countless other topics. There is an important geographical dimension to virtually every subject, because virtually every subject is immersed in the context of the global environment. It is obviously necessary to understand geography in order to understand history (Napoleon's failure to conquer Russia cannot be understood apart from a knowledge of Russian geography), but it is equally important to understand geography in order to comprehend biology. The fact that lemurs evolved exclusively on Madagascar, for example, can only be explained with reference to geographic isolation, just as the singular existence of giant tortoises on Galapagos can only be explained in terms of geography. Geographic knowledge is equally indispensable in geology. It is impossible to discuss plate tectonics and continental drift without being familiar with the shapes and relative locations of the continents.

Second, fundamental geographic literacy is important because we live in a world that is increasingly interconnected and interdependent. The world's commodities and financial markets are inseparably intertwined in a global network of mutual dependence. What happens in London or Tokyo today will inevitably affect everyday life in the United States tomorrow. The world has always required soldiers and diplomats to have a global perspective; today it requires merchants, bankers, and financiers to do so as well. Professionals of every persuasion must know and understand the global connections that shape their lives if they hope to survive and prosper in a global environment. All of us need to be aware of the economic, cultural, political, environmental, and strategic relationships that link the various parts of the world to each other. Obviously, that kind of global awareness cannot be assembled in a vacuum. Fundamental geographic literacy is the essential foundation of a global perspective.

Third, fundamental geographic literacy is important because it provides a valuable context for constructing a broad and open mind. When people lack even a rudimentary geographic knowledge of the world in which they live—as do so many people in the United States—they are especially susceptible to the debilitating and limiting perspectives of provincialism and ethnocentrism. No one country is at the center of the world, and no one place is representative of the entire earth. Since making sound judgments about anything typically requires an effort at achieving some degree of objectivity, our judgments about the world will likely be improved if they are based on the kind of global perspective that stems from genuine geographic literacy.

This book is designed to help you acquire a basic knowledge of fundamental place-name geography. Completing the *Review Exercises* at the end of each chapter will enable you to build a mental map of the world

Places of the World

that will allow you to unhesitantly identify all of the world's principal geographic features, including each of the world's 194 independent countries. You'll want to obtain an up-to-date world atlas to use in conjunction with this book, but a current atlas and *Places of the World* will be all you'll need to achieve fundamental geographic literacy.

PLACES of the World

While being able to locate each of the world's 194 countries is necessary for developing a global perspective, it is hardly sufficient. Simply being able to name a country and find it on a map is not especially helpful if you know nothing at all about that country. In order to claim a useful knowledge of fundamental place-name geography, it is necessary to be familiar with a few basic facts about each country in the world. Therefore, in addition to some general background information about each world region, the chapters that follow will describe six fundamental geographic facts about each and every one of the world's 194 independent countries. Those six facts, which will allow you to sketch the essential profile of every major place-name location in the world, can be summarized by the acronym *PLACES*, which stands for *P*opulation, *L*anguage, *A*rea, *C*apital, *E*conomy, and *S*overeignty.

P is for *POPULATION*

An estimate of the current **total population** will be provided for each country, along with a calculation of the country's **population density**, expressed in terms of the number of people per square mile. There is enormous variance in population size among the world's independent countries, from a low of fewer than 1,000 residents of Vatican City to a high of more than 1.3 billion citizens of China (see *Appendix D* for a comparative listing of all the world's countries with regard to population). Worldwide population densities vary enormously as well, from a low of 5 people per square mile in Mongolia to a high of 44,000 people per square mile in Monaco. With a total population of 6.6 billion people and a total land area (excluding Antarctica) of approximately 52,500,000 square miles, the average population density of the earth's habitable land surface is about 126 people per square mile.

You may find it helpful to remember the total population and population density of an area with which you are familiar, such as your home state, as a useful frame of reference for comparing the relative sizes of other countries. The state of Florida, for example, has about 19 million residents and a population density of around 285 people per square mile. The United States of America, with a population of approximately 301 million, is one of the world's largest countries, but its population density as a whole is relatively low at only 79 people per square mile.

L is for *LANGUAGE(S)*

The official and/or **principal language** or languages will be provided for each country. The linguistic affiliations of a country's population are very telling and significant features, because linguistic boundaries delineate cultural boundaries and identify people who share common traditions, customs, beliefs, and values. When the former Soviet Union broke apart in

1991, for example, it fragmented along linguistic boundaries: today the fifteen independent countries that once comprised the U.S.S.R. have fifteen different official languages.

There are more than 5,000 languages spoken in the world, although many of the isolated and indigenous languages are rapidly dying out as the world becomes more and more culturally homogenized. Many countries in the world have more than one official or principal language; Papua New Guinea, where more than 700 languages are spoken, has the greatest linguistic diversity of any country on earth. English has become the world's *lingua franca*, meaning that it serves as an international medium of communication among speakers of different languages (especially in the areas of aviation, commerce, diplomacy, and science), and it is probably spoken by more people on earth than any other language (counting the large number of people throughout the world who speak English as a second language).

A is for *AREA*

The **total land area**, expressed in square miles, will be provided for each country. There is a huge range from the smallest to the largest country in the world: Vatican City is at one extreme, with less than 1/5th of one square mile, while Russia is at the other extreme, with nearly 6.6 million square miles stretched across eleven time zones (see *Appendix C* for a comparative listing of all the world's countries with regard to area).

Again, you may find it helpful to remember the total square mileage of some familiar place in order to be able to make meaningful comparisons with other countries around the world. Recalling our previous examples, the state of Florida has 65,758 square miles, and the United States as a whole has nearly 3.8 million square miles.

C is for *CAPITAL*

The **capital city** will be provided for each country. The capital city of any country is obviously the most important city in the nation on a political basis, but it is not necessarily the most important on an economic or cultural basis, nor is it necessarily the most populous city. The capital of the United States, for example, is a small city with little significance as a center of commerce, while the capital of France is the country's largest urban center and the focus of its economic and cultural life. Nevertheless, it is very useful to know and recognize the name of the capital city of every country in the world, especially given the news media's habit of referring to a country's government by its capital ("Moscow said today...").

E is for *ECONOMY*

The **per capita gross domestic product** will be provided for each country. The gross domestic product, or GDP, is the total value of all final goods and services produced within a nation in a given year. The *per capita* GDP is simply the average contribution each citizen makes to the gross domestic product—computed, of course, by dividing the GDP by the total population. (The U.S. dollar figures for per capita GDP presented in this book are derived from purchasing power parity [PPP] calculations rather than from conversions at official currency exchange rates.)

The per capita gross domestic product is intended to provide some fairly precise sense of the country's degree of affluence. There are, obviously, many other meaningful measures of a nation's economy that would merit examination (including per capita income, rate of inflation, currency exchange rate, and the like). Furthermore, per capita GDP cannot be interpreted as a direct or simple measure of a country's standard of living (many other factors would have to be considered for that, including literacy rates, longevity, crime rates, and so forth). Nevertheless, the per capita GDP provides a useful scale by which to compare the world's countries on a standard basis.

A worldwide comparison of per capita GDP's reveals an unsurprising but highly significant pattern: the countries of the developed world, including northern North America, northwestern Europe, and Japan, tend to have the world's highest figures; the countries of the underdeveloped world, especially Africa, exhibit the world's lowest figures (see *Appendix E* for a comparative listing of all the world's countries with regard to per capita GDP). The European country of Luxembourg has the highest per capita gross domestic product in the world, at $71,400; the African country of Malawi has the lowest per capita GDP, at $600.

S is for *SOVEREIGNTY*

The form of government and original date of independence will be provided for each of the world's 194 sovereign states. A political entity possesses **sovereignty** if it exercises independent and exclusive control over the internal and external affairs of a clearly defined territory and its population. ("Sovereign state" and "independent country" are used here as synonymous terms.) There are many different forms of government among the world's sovereign states, but most independent countries can be classified as either monarchies or republics.

A **monarchy** is a sovereign state with an hereditary head of state (i.e., a monarch such as a king, emperor; sultan, or emir); a **constitutional monarchy** is a state that places legal limits on the monarch's authority. A **republic** is a sovereign state with a head of state who is anything other than a monarch (e.g., a president, premier, or dictator). A **democratic republic** is a system of government in which supreme authority is vested in the people and exercised through a system of elected representation. A **socialist republic** is a totalitarian system of government in which a single authoritarian party controls the state-owned means of production.

Many of the countries in the Commonwealth (see *Appendix B*) are classified as **parliamentary democracies**, because by strict definition they are neither republics nor monarchies. All of the countries in the Commonwealth are fully sovereign in terms of their internal and external affairs, with an elected prime minister who serves as head of government (and are thus functionally republican), but most Commonwealth countries recognize the British monarch, Queen Elizabeth II, as their formal chief of state (and are thus nominally monarchical).

The majority of the world's independent countries are **unitary** states, meaning that they have a highly centralized form of political organization in which local governments owe their existence and authority to the national government. Local governments in unitary states lack the authority for policy initiatives; as a result, services such as education and health care are uniform throughout the country. (The democratic republic of France is an example of a unitary state.) Most of the world's larger and more ethnically-diverse countries are **federal** states, meaning that they have a legally-mandated division of authority between local governments and the national government. Local governments in federal states have the authority to initiate policy within certain bounds; as a result, services such as education and health care can vary from region to region within the country. (The democratic republic of the United States is an example of a federal state.)

The second half of the 20th century witnessed an enormous proliferation of independent countries in the world. Since 1950, when there were only 82 sovereign states on earth, the number of independent countries has more than doubled. The end of the era of colonialism has brought about a rising tide of independence, but the trend toward political separation and fragmentation has not been limited to the former colonial world: in the final decade of the 20th century, the three former countries of Czechoslovakia, the Soviet Union, and Yugoslavia split into twenty-two independent states (which became twenty-three in the first decade of the 21st century, when Serbia and Montenegro split into two countries).

The break-up of the former Soviet Union and the collapse of communism in eastern Europe in recent years suggest another worldwide political trend, namely the possibility that authoritarian governments may be on the decline. Yet at the same time the world is witnessing increasing sectarianism and nationalism, and civil wars around the globe are laying the foundations for opportunistic dictatorial regimes. The description of each country's sovereignty is intended to provide a political snapshot of the world as it exists at the moment; it should be kept in mind, however, that the political face of the planet is in a state of continual flux.

A Word about Quantitative Data

The quantitative information presented in this book has been compiled from a wide variety of current and authoritative sources, but all of it should be regarded as approximately correct rather than absolutely correct. It is extremely difficult if not impossible to assemble numerical data about the world's countries that are uniformly accurate. Many quantities, such as population figures, cannot be measured exactly but can only be estimated, and the reliability of such estimates varies widely. Moreover, many quantitative facts, such as per capita gross domestic product, are constantly changing, and some sources of information are inevitably more current than others.

There is an additional difficulty in obtaining accurate numerical information in the realm of world geography: many presumably authoritative sources contradict one another, even about quantitative facts that are neither especially difficult to measure nor frequently changing. For example, China is described in many reference sources as slightly larger in land area than the United States, but just as many sources describe the U.S.A. as slightly larger than China. Even those sources that agree on the *relative* size of the two countries do not always agree on their *absolute* sizes. (Most of the quantitative information presented in this book is derived from either *Goode's World Atlas* or *The World Factbook*, and those two sources are generally congruent with one another—both list the United States as slightly larger than China, for example.)

In the final analysis, all of the numerical data presented in this book should be regarded as estimates and approximations. The quantitative data contained in these pages will allow meaningful and generally accurate comparisons between countries, but they should not be used for specific, narrowly focused comparisons where very precise distinctions are sought.

A Note about Pronunciation

Throughout this book, pronunciation guides have been provided for many of the geographic place-names with which American students are likely to be less familiar. Those guides will prove useful in formulating pronunciations that will be widely recognized according to the conventions of American English, but they should not be regarded as definitive. Various dialects of English (such as Australian English, British English, South African English, and West Indian English) often have widely varying pronunciations for the same place-names. Even within a single dialect, there is often little consensus regarding the "correct" pronunciation of a given name. Different atlases, dictionaries, and encyclopedias published in the United States, for example, often specify different ways of pronouncing the same geographic locations. If you follow the guides in this book, however, you will be able to pronounce every place-name in the world in a manner that will be immediately recognizable to speakers of American English.

The pronunciation guides in this book use an informal phonetic system that will allow you to "sound out" the more exotic place-names. The conventional rules of English spelling have been used when those rules would result in a clear and unambiguous pronunciation (under those rules, for example, the word "train" could be spelled *trane*). In other cases, the principal vowel sounds have been transcribed as follows:

"a" *as in* "art"	=	ah
"a" *as in* "way"	=	ay
"e" *as in* "ebb"	=	eh
"e" *as in* "equal"	=	ee
"i" *as in* "if"	=	ih
"i" *as in* "ice"	=	eye *or* igh
"o" *as in* "over"	=	oh
"o" *as in* "ooze"	=	oo
"o" *as in* "out"	=	au
"u" *as in* "up"	=	uh

Some vowel sounds have been variously transcribed with the goal of making the pronunciation immediately apparent in various contexts. The long "i" sound, for example, has been transcribed as "eye" in those cases where the vowel sound constitutes an entire syllable, as in the Iberian (eye-BIHR-ee-uhn) Peninsula; where the long "i" sound constitutes only part of a syllable, however, it has been transcribed as "igh" to rhyme with the word *sigh*, as in the Sinai (SIGH-nigh) Peninsula.

INTRODUCTION
Planet Earth

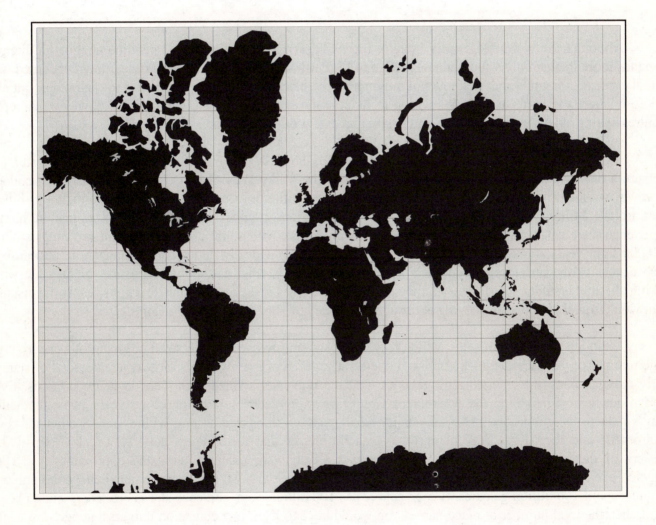

Planet Earth

Although the word **geography** literally means "description of the earth," genuine geographic literacy demands more than simply a focused knowledge of the earth itself. Since geography is centrally concerned with spatial context, it is important to know not only where the places on earth are in relation to each other, but also to know where the earth itself is in relation to its celestial neighbors. Basic geographic knowledge therefore requires having at least a general sense of the size and extent of the universe.

As most American students learn in grade school, the earth is one of eight planets orbiting the sun, and the sun is one of a large number of stars that make up the Milky Way Galaxy. Many students reach college, however, without an informed appreciation of just how vast the universe is. There are more than 100 billion stars in the Milky Way Galaxy, for example--an enormous number that dwarfs most things in human experience. Our galaxy, which is shaped like a flattened disk with spiral arms that extend outward, is about 100,000 light years across--again, an enormous distance that practically defies conceptualization. (Because of the vastness of the universe, astronomers avoid using miles or other earthbound measures of distance when talking about interstellar space. Instead, they use **light years** as a measure of distance: a light year is the distance light would travel in one year, moving at the constant speed of 186,283 miles per second.)

Although it is enormous, the Milky Way Galaxy is simply one average-sized grouping of stars among a huge number of galaxies that make up the universe. The force of gravity keeps most concentrations of stars in fairly close proximity to neighboring galaxies, and the universe is filled with galactic neighborhoods. Even though our nearest neighbor, the Andromeda Galaxy, is an unreachable 2 million light years away, the Milky Way's galactic neighborhood is actually fairly small, as such things go. Along with some 18 other galaxies, Andromeda and the Milky Way comprise a galactic cluster that astronomers call the Local Group. Other clusters of galaxies are much larger. The Hercules Cluster, for example, contains approximately 10,000 galaxies, each containing on average approximately 100 billion stars. All told, astronomers believe the universe contains about 100 billion galaxies. If each galaxy has 100 billion stars, and each star is orbited by eight planets or so, then the total number of place-names you would have to learn in a course on universal geography would be truly astronomical.

Fortunately, the goal of this book is much more modest. For our purposes, we will focus on acquiring a fundamental familiarity with the planet earth, and ignore for the most part the unimaginably vast universe that stretches out beyond our atmosphere. Even the most modest level of geographic literacy, however, cannot entirely ignore the larger context of spatial relations. For that reason, you should become at least generally familiar with the earth's immediate neighborhood.

The Solar System

The eight planets of the solar system are listed below in order of their distance from the sun, beginning with the planet closest to the sun; each planet's size rank is indicated in parentheses following the planet's name.

Planet	Average Distance from Sun		Diameter *(in miles)*	Average Density *(water = 1)*	Satellites *(number)*
	(in miles)	*(in light years)*			
Mercury (8)	35,990,000	3.2 minutes	3,032	5.5	0
Venus (6)	67,240,000	6.0 minutes	7,523	5.3	0
Earth (5)	93,000,000	8.3 minutes	7,926	5.5	1
Mars (7)	141,730,000	12.7 minutes	4,220	4.0	2
Jupiter (1)	483,880,000	43.3 minutes	88,750	1.3	16
Saturn (2)	887,130,000	1.3 hours	74,580	0.7	23
Uranus (3)	1,783,700,000	2.7 hours	31,600	1.3	15
Neptune (4)	2,795,500,000	4.2 hours	30,200	1.8	8

In 2006, the International Astronomical Union passed a resolution redefining a "planet" as a round celestial body orbiting the sun that has cleared the neighborhood around its orbit. Because Pluto has not cleared the neighborhood around its orbit, it is no longer considered a planet (compared to the eight planets, Pluto also has a highly-elliptical and oddly-angled orbit, and it is smaller than several other moons in the Solar System, including the Earth's—Pluto is now classified as a "dwarf planet").

Geographic Terms

We turn now to the central geographic task of describing the earth, and we need to begin with a brief vocabulary lesson. The following list contains a number of basic geographic terms with which you should be familiar. Unfortunately, not all of these terms are used consistently in the world's various place-names, and not all of them are defined unambiguously. The Caspian Sea, for example, is not really a sea at all, while Australia is sometimes designated as an island, sometimes as a continent, and sometimes as both. Nevertheless, it is important to have a basic understanding of the terms geographers use to describe the earth's physical features.

Archipelago An archipelago is a collection of closely grouped islands. With more than 13,000 islands, **Indonesia** is the world's largest archipelago country (*pronounced* AHR-kuh-PEHL-uh-GOH).

Bay A bay is a relatively small arm of a body of water that penetrates into a surrounding landmass. In general, bays are smaller than gulfs (**Chesapeake Bay** and **San Francisco Bay** are typical examples), but there are many exceptions (the **Bay of Biscay**, for instance, is substantially larger than the **Gulf of Aqaba**).

Canal A canal is an artificial **channel** or **strait**. The **Panama Canal** and the **Suez Canal** are two of the world's most important constructed channels.

Cape A cape is a point of land that extends into a body of water. A cape could be formed by an arm of the mainland (such as **Cape Agulhas** in South Africa, which forms the southern terminus of the African continent), or it could be formed by an extension of an island (such as **Cape Horn** in the Tierra del Fuego archipelago, which forms the southern terminus of the South American continent).

Channel A channel is a wide strip of water that separates two nearby landmasses and connects two larger bodies of water. The **English Channel**, for example, separates the island of Great Britain from the European continent and connects the Atlantic Ocean to the North Sea.

Continent A continent is a relatively large body of land that generally serves to define the boundaries between oceans. Traditionally, the world is divided into seven continents: North America, South America, Europe, Asia, Africa, Australia, and Antarctica. Strictly speaking, all of the continents could be called islands, because all of the continental landmasses are entirely surrounded by water (even though North and South America form one contiguous landmass, as do Europe, Asia, and Africa). The fact that the term "island" is generally reserved for smaller landmasses, however, makes the status of **Australia** somewhat ambiguous. It is the world's smallest continent, but is it also the world's largest island? The answer frequently varies from one geography textbook to another. For our purposes, we will call Australia a continent and identify **Greenland** as the world's largest island.

Estuary The area at the mouth or lower course of a river where the river's current meets the sea's tide, mixing freshwater and saltwater.

Fjord A long narrow arm of the sea bordered by steep cliffs, typically formed by glacial erosion of river valleys; coastlines that are intercut with many fjords, such as Norway's, are extremely irregular.

Gulf A gulf is a relatively large arm of a body of water that penetrates into a surrounding landmass. In general, gulfs are larger than bays (the **Gulf of Mexico** is an outstanding example), but there are many exceptions (the **Gulf of Suez**, for instance, is substantially smaller than **Hudson Bay**).

Island An island is a relatively small body of land that is entirely surrounded by water. With an area of a little more than 300,000 square miles, **New Guinea** is the world's second largest island. *See* **continent**.

Isthmus An isthmus is a narrow strip of land that separates two nearby bodies of water and connects two larger bodies of land. The **Isthmus of Panama**, for example, separates the Pacific Ocean from the Atlantic Ocean and connects North America to South America.

Lake A lake is a body of water (either fresh or salt) that is entirely surrounded by land and which has no natural outlet to the world ocean (**Lake Baikal**, for example, is the world's deepest lake, and

Lake Titicaca is the world's highest navigable lake). The **Caspian Sea** is the world's largest lake, despite its name, because it has no natural connection to the ocean. The Caspian Sea is, however, salty; it even has a species of seal that is found nowhere else in the world.

Ocean An ocean is an extremely large body of water whose borders are defined by continental landmasses. The world is considered to have five major oceans: **Pacific**, **Atlantic**, **Indian**, **Southern**, and **Arctic**. In fact, however, there is only one world ocean, because all five customary divisions of the world ocean are connected to each other. Indeed, the dividing points between the five oceans are not always defined with great precision.

Peninsula A peninsula is a piece of land that projects into a body of water and is surrounded by water on three sides. In general, a peninsula extends farther into the water than a cape, which typically has only two sides. **Baja California** is a classic example of a peninsula.

Sea A sea is a relatively large portion of the ocean that is partially shaped by landmasses, although the boundary between a sea and its adjacent ocean is often not well defined (the boundary between the **Arabian Sea** and the **Indian Ocean** is a good example). Other seas, however, may be clearly demarcated by surrounding land forms (the **Mediterranean Sea** is an obvious example). A sea is generally larger than a gulf or a bay (the **Caribbean Sea**, for example, is larger than most bodies of water that are called gulfs or bays), but again there are exceptions; the **Bay of Bengal**, for instance, is much larger than the **Baltic Sea**.

Strait A strait is a narrow strip of water that separates two nearby landmasses and connects two larger bodies of water. The **Strait of Gibraltar**, for example, separates Spain from Morocco and connects the Atlantic Ocean to the Mediterranean Sea.

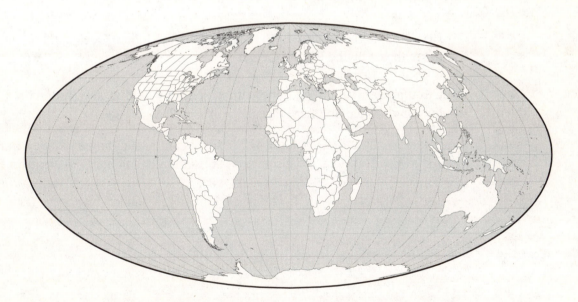

Major Surface Features of the Earth

The locations of most of the major features of the earth's surface are illustrated with black-and-white schematic maps on the following pages. In order to build a vivid mental image of these various land forms and bodies of water, however, you should consult a full-color, detailed world atlas to help you remember the precise dimensions and locations of each feature.

Continents

North America

Bordered on the west by the Pacific Ocean, on the north by the Arctic Ocean, on the east by the Atlantic Ocean, and terminating in the south at the Panama/Colombia border

South America

Bordered on the west by the Pacific Ocean, on the north by the Caribbean Sea and the Panama/Colombia border, on the east by the Atlantic Ocean, and terminating in the south at Cape Horn

Europe

Bordered on the west by the Atlantic Ocean, on the north by the Arctic Ocean, on the south by the Mediterranean Sea, and on the east by the Ural Mountains, Caspian Sea, Caucasus Mountains, and Black Sea

Asia

Bordered on the north by the Arctic Ocean, on the east by the Pacific Ocean, on the south by the Indian Ocean, and on the west by the Ural Mountains, Caspian Sea, Caucasus Mountains, Black Sea, Mediterranean Sea, and Red Sea

Africa

Bordered on the west by the Atlantic Ocean, on the north by the Mediterranean Sea, on the east by the Red Sea and the Indian Ocean, and terminating in the south at Cape Agulhas

Australia

Bordered on the southwest to northwest by the Indian Ocean and on the northeast to southeast by the Pacific Ocean

Antarctica

Circumpolar landmass lying almost entirely within the Antarctic Circle and bordered by the Southern Ocean

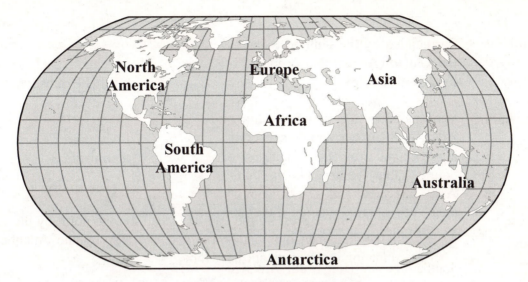

Capes and Peninsulas

Florida Peninsula
Surrounded by the Atlantic Ocean and the Gulf of Mexico, and occupied by the U.S. state of the same name

Baja California
(BAH-hah)
Western Mexico, surrounded by the Gulf of California and the Pacific Ocean

Yucatan Peninsula
(YOO-kah-tahn)
Surrounded by the Gulf of Mexico and the Caribbean Sea, and occupied by the countries of Mexico, Belize, and Guatemala; separated from Cuba by the Yucatan Channel

Iberian Peninsula
(eye-BIHR-ee-uhn)
Southwestern Europe, occupied by the countries of Portugal and Spain

Italian Peninsula
Southern Europe, occupied by the countries of Italy, San Marino, and Vatican City

Scandinavian Peninsula
Northern Europe, occupied by the countries of Norway and Sweden

Balkan Peninsula
Southeastern Europe, occupied by the countries of Slovenia, Croatia, Bosnia-Herzegovina, Serbia & Montenegro, Macedonia, Albania, Greece, Bulgaria, and Romania

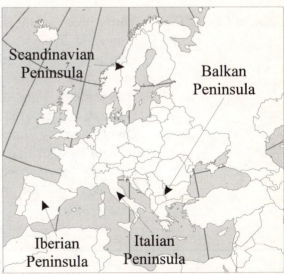

Sinai Peninsula (SIGH-nigh)	Surrounded by the Gulf of Suez and the Gulf of Aqaba, and occupied by the country of Egypt	
Arabian Peninsula	Surrounded by the Red Sea, Gulf of Aden, Arabian Sea, and Persian Gulf , and occupied by the countries of Saudi Arabia, Yemen, Oman, United Arab Emirates, and Qatar; with an area of approximately 1,000,000 square miles, it's about 1½ times the size of Alaska	

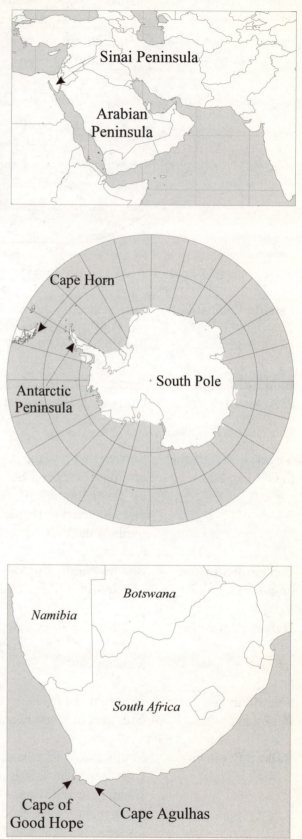

Cape Horn	Southernmost tip of South America in the Chilean portion of the Tierra del Fuego archipelago (divides Pacific and Atlantic Oceans at approximately 67° West Longitude); Dutch navigator Willem Schouten named it in 1616 for his birthplace, the Dutch city of Hoorn
Antarctic Peninsula	Antarctica, south-southeast of Tierra del Fuego in South America; most of the land in Antarctica lies inside the Antarctic Circle—the 700-mile long Antarctic Peninsula is the most prominent exception (the peninsula's northern part is known as Graham Land, the southern as Palmer Land)
Cape of Good Hope	Thirty miles south of Cape Town in South Africa's Western Cape Province; named the "Cape of Storms" by the Portuguese navigator Bartolomeu Dias in 1488, and subsequently renamed the "Cape of Good Hope" by Portugal's King John II, who thought that discovering the extremity of the African continent augured well for future maritime enterprises

Cape Agulhas
(ah-GULL-ahss)

Southernmost point on the African Continent; at 20° degrees east longitude, it sits on the meridian that divides the Atlantic and Indian Oceans

Malay Peninsula
(mah-LAY)

Southeast Asia, surrounded by the Bay of Bengal, the Strait of Malacca, and the South China Sea, and occupied by the countries of Burma, Thailand, and Malaysia (the island nation of Singapore lies immediately off the southern tip of the peninsula); a mountain range extends the entire length of the peninsula

Korean Peninsula

Northeast Asia, bordered on the west by the Yellow Sea and on the east by the Sea of Japan, and occupied by the countries of North Korea and South Korea (which have been divided from each other along the 38th parallel since 1948); on the north, the peninsula is bordered by China (the northeast corner of North Korea is bordered by Russia); the chain of islands along the peninsula's southern coast is the Korean Archipelago

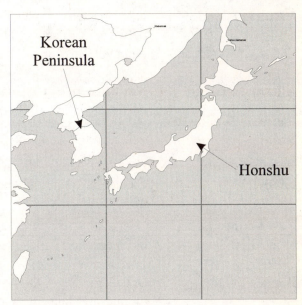

Islands

The locations of the twenty largest islands in the world are described and illustrated below. All but three of these islands are larger than 40,000 square miles in area (each island's size rank is given in parentheses following its name).

Greenland (1st)   Northeast of Canada; a political dependency of Denmark

Baffin Island (5th)   Northeastern Canada, guarding the entrance to Hudson Bay; part of Canada's Nunavut Territory

Victoria Island (9th)   Northwestern Canada; divided between Canada's Northwest and Nunavut Territories

Ellesmere Island (10th)   Northeastern Canada, off the northwest coast of Greenland; part of Canada's Nunavut Territory

Newfoundland Island (16th)   Eastern Canada; the insular part of Canada's Newfoundland Province

New Guinea (2nd) (new GIHN-ee)   Occupied by the countries of Indonesia and Papua New Guinea

Borneo (3rd)   Occupied by the countries of Indonesia, Malaysia, and Brunei

Sumatra (6th) (soo-MAH-truh)   Occupied by the country of Indonesia; separated from the Malay Peninsula by the Strait of Malacca

Celebes (11th) (SELL-uh-beez)   Occupied by the country of Indonesia; also known as Sulawesi

Java (13th)   Occupied by the country of Indonesia, and home to its capital city of Jakarta

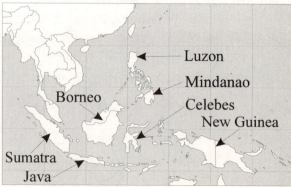

Luzon (17th)
(loo-ZAHN)

Largest island in the Philippines archipelago, home to its capital city of Manila

Mindanao (19th)
(min-dah-NOW)

Second largest island in the Philippines archipelago

South Island (12th)

Larger of the two main islands that comprise the country of New Zealand; the mountain range known as the Southern Alps is located on South Island—its highest peak is Mt. Cook, at more than 12,000 feet above sea level

North Island (14th)

Smaller of the two main islands that comprise the country of New Zealand; it's also mountainous, with an irregular coastline that features excellent harbors

Honshu (7th)
(honn-SHOO)

Largest of the four main islands that comprise the country of Japan, and home to its capital city of Tokyo; the other three islands are Hokkaido (north of Honshu), Shikoku, and Kyushu; Honshu is considered the mainland of Japan—with an area of approximately 86,000 square miles, it's about the size of Minnesota

Madagascar (4th)
(mad-uh-GAS-kar)

Large island off the southeast coast of Africa, occupied by the country of the same name; nearly 1,000 miles long, it's separated from the African mainland by the Mozambique Channel; plateaus and mountains cover the island (the highest peak is nearly 9,500 feet above sea level)

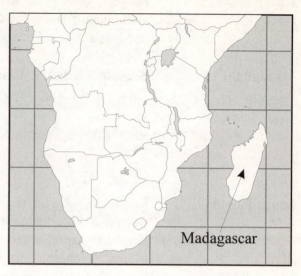

Cuba (15th)

Largest island in the Caribbean, occupied by the country of the same name; nearly 750 miles long, it's separated from the North American mainland by the Straits of Florida; its 2,000-mile coastline has numerous cays and islands (the largest of which is Isla de la Juventud off the southwest coast)

Great Britain (8th)   Largest island in the British Isles, and the largest island in Europe, it's occupied by the country of United Kingdom (specifically, the English, Scottish, and Welsh portions of the UK); with just over 88,000 square miles, it's slightly larger than Minnesota (or nearly twice the size of Ohio)

Ireland (20th)   Second largest of the British Isles, separated from Great Britain by the Irish Sea and St. George's Channel; occupied by the Republic of Ireland in the south, central, and northwestern parts, and the United Kingdom's Northern Ireland in the northeast; with an area of approximately 32,000 square miles, it's about the size of South Carolina

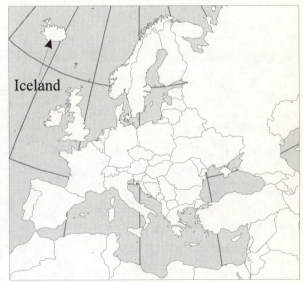

Iceland (18th)   North Atlantic island between Greenland and Norway, lying immediately south of the Arctic Circle and occupied by the country of the same name; approximately 40,000 square miles in area, it's about the size of Kentucky, with a 3,700-mile long coastline that's indented with many long fjords; with more than 100 volcanoes and 120 glaciers, its nickname is the "Land of Fire and Ice"

Bodies of Water

Pacific Ocean   Bordered on the west by Australia and Asia, on the north by the Arctic Ocean, on the east by North and South America, and on the south by the Southern Ocean; divided by the Equator into the North and South Pacific

Atlantic Ocean   Bordered on the west by North and South America, on the north by the Arctic Ocean, on the east by Europe, Asia, and Africa, and on the south by the Southern Ocean; divided by the Equator into the North and South Atlantic

Indian Ocean   Bordered on the west by Africa, on the north by Asia, on the east by Australia, and on the south by the Southern Ocean

Southern Ocean   Extends from the coast of Antarctica north to 60° South Latitude; consisting of what were formerly the southern portions of the Pacific, Atlantic, and Indian Oceans, the Southern Ocean was delimited in 2000 by the International Hydrographic Organization

Arctic Ocean   Circumpolar area largely within the Arctic Circle bordered by North America, Europe, Asia, and the Atlantic and Pacific Oceans

Mediterranean Sea Eastern arm of the Atlantic Ocean, bordered on the west by the Strait of Gibraltar, on the north by Europe, on the east by Asia, on the south by Africa; connected to the Red Sea via the Suez Canal, and the Black Sea via the Dardanelles, Sea of Marmara, and Bosporus; with an area of about 1,000,000 square miles, it's virtually tied with the Caribbean for the title of largest sea in the world

Tyrrhenian Sea (tuh-REE-nee-an) Bordered on the west by Sardinia, on the east by Italy, on the south by Sicily; the Bay of Naples is an inlet of the Tyrrhenian Sea

Adriatic Sea Bordered on the west by Italy and on the east by Slovenia, Croatia, Bosnia, Serbia & Montenegro, and Albania

Ionian Sea (eye-OWN-ee-uhn) Bordered on the west by Italy and on the east by Greece

Aegean Sea (uh-JEE-un) Bordered on the west by Greece and on the east by Turkey; includes numerous islands in several groups

Black Sea Bordered on the west by Romania and Bulgaria, on the north by Ukraine, on the east by Russia and Georgia, on the south by Turkey

Dardanelles (dahr-duh-NELZ) Strait connecting the Aegean Sea and the Sea of Marmara

Sea of Marmara Bordered on the west by the Dardanelles, on the east by the Bosporus, on the north and south by Turkey

Bosporus (BAHS-puhr-us) Strait connecting the Black Sea and the Sea of Marmara

Bay of Biscay
(bihss-KAY)

Large inlet of the Atlantic Ocean on the west coast of France and the north coast of Spain; also known as the Gulf of Gascony

Strait of Gibraltar

Separating Europe (Spain and Gibraltar) from Africa (Morocco and Ceuta) and connecting the Atlantic Ocean and Mediterranean Sea; 36 miles long, it varies from 8 to 23 miles wide

English Channel

Separating United Kingdom from France and Belgium and connecting the Atlantic Ocean and the North Sea (via the Dover Strait); the French call the channel *La Manche*

North Sea

Bordered on the west by Great Britain, on the north by the Shetland Islands, on the east by Norway and Denmark, on the south by Netherlands and Germany; also known as the German Ocean, it's the site of extensive oil and natural gas fields

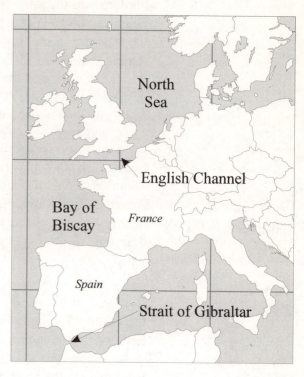

Baltic Sea

Bordered on the west by Denmark and Sweden, on the east by Finland, Russia, Estonia, Latvia, and Lithuania, on the south by Poland and Germany; connected to the North Sea by a series of straits (Skaggerak, Kattegat, and Øresund) that encircle Denmark; with an area of some 163,000 square miles, it's about the size of California

Gulf of Bothnia

Northern arm of the Baltic Sea between Sweden and Finland

Gulf of Finland

Eastern arm of the Baltic Sea between Finland, Russia, and Estonia

Norwegian Sea

Northernmost arm of the Atlantic Ocean, bordered on the west by Iceland and the Greenland Sea, on the north by the Arctic Ocean, on the east by Norway, on the south by the Faeroe Islands; in the Norwegian Atlantic Current, a branch of the Gulf Stream carries warm water northward and contributes to the relatively mild and moist climate of Norway

Caspian Sea

Bordered on the west by Russia and Azerbaijan, on the east by Kazakhstan & Turkmenistan, on the south by Iran; with nearly 144,000 square miles (about the size of Montana), the Caspian Sea is the world's largest lake; it serves as part of the border between Europe and Asia

Aral Sea
(uh-RAHL)

Bordered on the northwest to northeast by Kazakhstan and on the southwest to southeast by Uzbekistan

Lake Baikal
(bigh-KAHL)

Eastern Russia (Siberia), north of central Mongolia; nearly 12,000 square miles in area, it's about the size of Maryland; with a maximum depth of 5,315 feet, it's the deepest lake in the world, and the only one that's more than one mile deep; also spelled "Baykal"

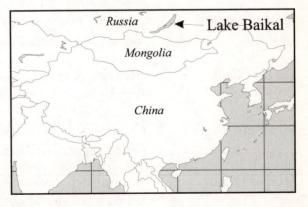

Arabian Sea

Arm of the Indian Ocean bordered on the west by the Gulf of Aden and the Arabian Peninsula (specifically the countries of Oman and Yemen), on the northwest by the Gulf of Oman, on the north by Pakistan, on the east by India; except for the relatively narrow subtropical region along the Pakistani coast, the Arabian Sea is entirely tropical

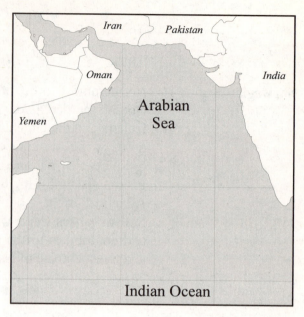

Red Sea

Bordered on the west by Africa and on the east by the Arabian Peninsula; connected to the Mediterranean via the Suez Canal, and to the Gulf of Aden via the Bab el Mandeb

Gulf of Suez

Extreme northwestern arm of the Red Sea; city of Suez is on the north coast

Gulf of Aqaba
(AH-kah-buh)

Extreme northeastern arm of the Red Sea; city of Aqaba is on the north coast

Suez Canal

Dividing the rest of Egypt from the Sinai Peninsula and connecting the Mediterranean Sea & the Gulf of Suez

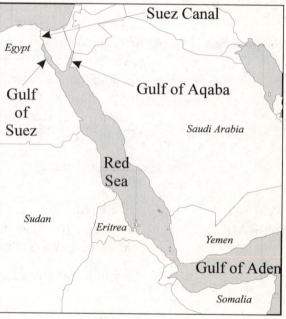

Gulf of Aden

Arm of the Arabian Sea between Yemen and Somalia

Persian Gulf

Between Arabian Peninsula, Kuwait, Iraq, and Iran, a.k.a. "The Gulf"

Strait of Hormuz

Connecting Persian Gulf to Gulf of Oman, separating Iran and Oman

Gulf of Oman

Northwestern arm of the Arabian Sea between Oman and Iran

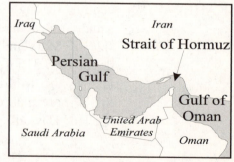

18   Places of the World

Bay of Bengal
(benn-GAHL)

Bordered on the west by India and Sri Lanka, on the north by Bangladesh, and on the east by Burma; the Ganges River, whose mouths form a wide delta in the border region of India and Bangladesh, flows into the Bay of Bengal; in December 2004, the Bay of Bengal was the scene of devastating destruction from tsunamis created by an immensely powerful undersea earthquake whose epicenter was off the northwest coast of Sumatra

Yellow Sea

Bordered on the west by China and on the east by the Korean Peninsula; exceptionally shallow, with an average depth of 121 feet and a maximum depth of 250 feet

Sea of Japan

Bordered on the west by the Korean Peninsula and on the east by Japan; connected to the East China Sea via the Korea Strait, which separates South Korea from Japan

East China Sea

Bordered on the west by China, on the north by the southern extreme of the Korean Peninsula, on the east by the Ryukyu Islands, on the south by Taiwan

South China Sea	Bordered on the west by Vietnam and the Malay Peninsula, on the north by China and Taiwan, on the east by Philippines, on the south by Malaysia and Brunei

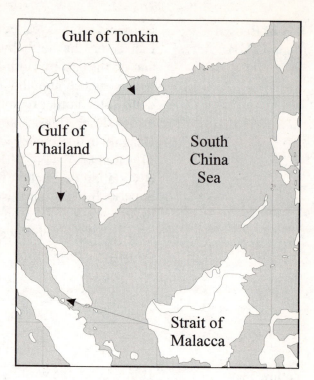

Gulf of Tonkin	Arm of the South China Sea encircled by Vietnam and China's Hainan Island; site of alleged Vietnamese firing on U.S. ships in 1964 leading to wider American military involvement in Vietnam
Gulf of Thailand	Arm of the South China Sea bordered on the west by the Malay Peninsula, on the north by Thailand, on the east by Cambodia
Strait of Malacca	Separating Sumatra from the Malay Peninsula and connecting the Indian Ocean to the Pacific Ocean; about 500 miles long, it varies in width from 40 miles to 300 miles
Timor Sea	Between northwest coast of Australia and the island of Timor in the Indonesian archipelago
Arafura Sea	Between north-central Australia and the western half of the island of New Guinea

Coral Sea	Bordered on the west by Australia, on the north and east by the islands of Melanesia, on the south by New Caledonia
Tasman Sea	Portion of the South Pacific bordered on the west by Australia and on the east by New Zealand; the Australian island of Tasmania lies on the sea's southwest border

Lake Victoria

East-Central Africa, surrounded by the countries of Uganda, Kenya, and Tanzania; roughly 250 miles long by 200 miles wide, it's about the size of Louisiana; Victoria is the second-largest freshwater lake in the world, after North America's Lake Superior—its only outlet is the Nile River

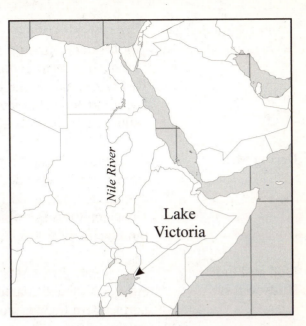

Lake Titicaca
(tee-tee-KAH-kah)

In the Andes Mountains, in west-central South America, surrounded by the countries of Peru and Bolivia; 120 miles long by 45 miles wide, it's about twice the size of Rhode Island; Titicaca sits at an elevation of 12,500 feet above sea level, making it the highest navigable lake in the world

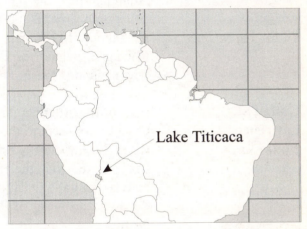

Drake Passage

Between the southernmost tip of South America and the Antarctic Peninsula (connects the Pacific and Atlantic Oceans); named for Sir Francis Drake, who from 1577 to 1580 became the first Englishman to circumnavigate the world

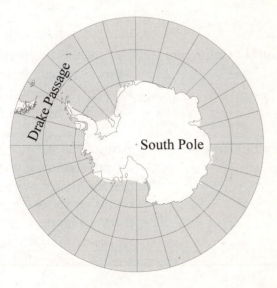

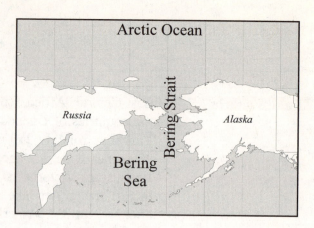

Bering Sea

Northernmost portion of the Pacific Ocean, bordered on the west by Russia, on the north by the Bering Strait, on the east by Alaska, on the south by the Aleutian Islands

Bering Strait

Separating Alaska from Russia and connecting the Pacific Ocean and the Arctic Ocean; named for the Danish navigator Vitus Bering, who traversed the strait in 1728

Hudson Bay

Arm of the Atlantic Ocean in northeastern Canada, connected to the Atlantic by the Hudson Strait; covering nearly 500,000 square miles, it's about five times the size of Michigan; named for Henry Hudson, who navigated the east coast in 1610

Gulf of St. Lawrence

Arm of the Atlantic Ocean in eastern Canada; connects to the St. Lawrence River in the northwest corner, and from there, via the St. Lawrence Seaway, to Lake Ontario; includes the Canadian province of Prince Edward Island

Gulf of California

Arm of the Pacific Ocean in western Mexico (also known as the Sea of Cortez, formerly known as the Vermillion Sea)

Gulf of Mexico

Arm of the Atlantic Ocean, bordered on the southwest to east by Mexico and the United States of America and on the southeast by a line connecting the Yucatan Peninsula and the Florida Keys; at 600,000 square miles, it's slightly smaller than Alaska

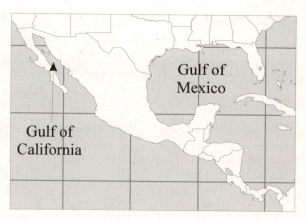

Caribbean Sea Arm of the Atlantic Ocean, bordered on the west by the Yucatan Peninsula and Central America, on the north by the Greater Antilles, on the east by the Lesser Antilles, on the south by Colombia and Venezuela; the islands enclosing the Caribbean—the Greater Antilles and the Lesser Antilles—are collectively referred to as the West Indies

Lake Superior On the border between Canada (Ontario) and the U.S.A. (Minnesota, Wisconsin, Michigan)

Lake Huron On the border between Canada (Ontario) and the U.S.A. (Michigan)

Lake Michigan Bordered by Wisconsin, Michigan, Indiana, Illinois

Lake Erie On the border between Canada (Ontario) and the U.S.A. (Michigan, Ohio, Pennsylvania, New York)

Lake Ontario On the border between Canada (Ontario) and the U.S.A. (New York)

Panama Canal Dividing the isthmus of Panama and connecting the Atlantic Ocean and the Pacific Ocean, it runs northwest to southeast across the Isthmus of Panama—the Atlantic (Caribbean) entrance to the canal is 27 miles *west* of the Pacific entrance; opened in 1914, the U.S. ceded control of the canal to Panama in 2000

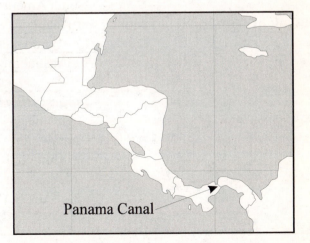

Dimensions of the Earth

It is important to have a fairly accurate sense of the size of the planet, so you should remember that the earth has a **diameter** of approximately 8,000 miles and a **circumference** of approximately 25,000 miles.

For the sake of precision, you might be interested to know that the equatorial diameter of the earth is 7,926.4 miles while the polar diameter is only 7,899.8 miles. The equatorial bulge is caused by the rotation of the earth about its axis, which means that the equatorial circumference is also slightly larger than the polar circumference (24,902 miles compared to 24,860 miles).

The following figures summarize the approximate sizes of the earth's principal surface features:

Surface	Area in Miles^2
Total Surface	197,000,000
Water Surface	139,100,000
Land Surface	57,900,000
Oceans	**Area in Miles^2**
Pacific Ocean	60,300,000
Atlantic Ocean	29,600,000
Indian Ocean	26,500,000
Southern Ocean	7,800,000
Arctic Ocean	5,400,000
Continents	**Area in Miles^2**
Asia	17,300,000
Africa	11,700,000
North America	9,500,000
South America	6,900,000
Antarctica	5,400,000
Europe	3,800,000
Australia	3,000,000
Largest Islands	**Area in Miles^2**
Greenland	840,000
New Guinea	305,000
Borneo	290,000
Madagascar	227,000
Baffin	196,000

The World's Largest and Smallest Countries

The chapters that follow will review the geographic and demographic sizes of every country in the world. At this stage, however, it will be useful to be able to identify the world's largest and smallest countries in order to appreciate the range of variation among the world's 194 sovereign states. Refer to your atlas to be sure that you can locate each of the following countries on a map of the world.

The Ten Largest Countries

	By Area (in square miles)			By Population	
1	Russia	6,592,849	1	China	1,321,852,000
2	Canada	3,855,103	2	India	1,129,866,000
3	USA	3,794,083	3	USA	301,140,000
4	China	3,690,045	4	Indonesia	234,694,000
5	Brazil	3,300,172	5	Brazil	190,011,000
6	Australia	2,969,910	6	Pakistan	164,742,000
7	India	1,222,510	7	Bangladesh	150,448,000
8	Argentina	1,073,519	8	Russia	141,378,000
9	Kazakhstan	1,049,156	9	Nigeria	135,031,000
10	Sudan	967,500	10	Japan	127,433,000

Notice that third-place USA and fifth-place Brazil are the only two countries that appear in the same position on both lists. Canada, Australia, Argentina, Kazakhstan, and Sudan all have relatively small populations and hence relatively low population densities. China, India, Pakistan, and Bangladesh form one contiguous landmass that contains by far the highest concentration of human habitation on earth.

The Ten Smallest Countries

	By Area (in square miles)			By Population	
1	Vatican City	.17	1	Vatican City	820
2	Monaco	.75	2	Tuvalu	12,000
3	Nauru	8	3	Nauru	14,000
4	Tuvalu	10	4	Palau	21,000
5	San Marino	24	5	San Marino	30,000
6	Liechtenstein	62	6	Monaco	33,000
7	Marshall Islands	70	7	Liechtenstein	34,000
8	St. Kitts & Nevis	101	8	St. Kitts & Nevis	39,000
9	Maldives	115	9	Marshall Islands	62,000
10	Malta	122	10	Antigua & Barbuda	69,000

Vatican City, Monaco, San Marino, Liechtenstein, Andorra, and Malta are all located in Europe, while Nauru, Tuvalu, and Palau are all located in the southwest Pacific. St. Kitts & Nevis and Antigua & Barbuda are both Caribbean island nations; the Maldives archipelago can be found in the Indian Ocean, just southwest of India.

Mapping the Earth

Geographers use a mathematical system of intersecting lines called the geographic grid to describe the exact location of any place on the surface of the earth. The **geographic grid** consists of two sets of imaginary lines that cross each other at 90 degree angles. One set of lines, called **parallels**, runs east-west; the other set, called **meridians**, runs north-south. Any location on the surface of the earth can be precisely plotted as the intersection of a particular parallel and a particular meridian.

The Geographic Grid

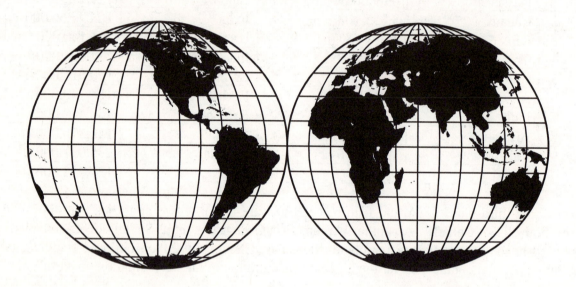

Parallels are also called lines of latitude. A line of **latitude** is a straight line that circles the globe in an east-west direction at right angles to the meridians. The parallel that provides the point of reference for all other lines of latitude is called the Equator. The **Equator** is the line of latitude that encircles the globe **equidistant** between the North and South Poles. The location of the Equator is thus determined by the natural conformation of the earth, as the **North Pole** and the **South Pole** mark the two end points of the earth's rotational axis. Since lines of latitude are parallel to each other, they are always an equal distance apart (just over 69 miles, which you can calculate by dividing the polar circumference of the earth, 24,860 miles, by the number of degrees in a circle, 360).

The Equator is designated the 0 degree line of latitude, and all other lines of latitude are measured either north or south of the Equator. If you go due north or due south from any point on the Equator, you will go exactly one-quarter of the way around the globe before coming to either the North or South Pole. Therefore the maximum number of degrees used to designate any line of latitude is 90 (the 360 degrees of a circle divided by 4). Washington, D.C., for example, is located at 38 degrees North Latitude, while Canberra, Australia is situated at 35 degrees South Latitude. The Equator divides the world into two equal halves: the **Northern Hemisphere**, the half of the earth that lies north of the Equator, and the **Southern Hemisphere**, the half of the earth that lies south.

Two other significant lines of latitude are determined by natural phenomena: the **Tropic of Cancer**, which is located at 23½ degrees North Latitude, and the **Tropic of Capricorn**, which is located at 23½ degrees South Latitude. The earth's rotational axis is not perpendicular to the plane of its orbit around the sun; instead, it is tilted 23½ degrees off the perpendicular. That phenomenon is responsible for the seasons, as well as for the location of the tropics. In June, when the North Pole is tilted toward the sun, it is summertime in the Northern Hemisphere and wintertime in the Southern Hemisphere. In December, when the South Pole is tilted toward the sun, it is wintertime in the Northern Hemisphere and summertime in the Southern Hemisphere. The Tropic of Cancer marks the northernmost point on earth where the sun's rays are directly overhead at noon; that event, called the **summer solstice**, occurs every year around June 21st. The Tropic of Capricorn marks the southernmost point on earth where the sun's rays are directly overhead at noon; that event, the **winter solstice**, occurs around December 22nd. Halfway between the two solstices, at the **spring** and **fall equinoxes** around March 21st and September 22nd, the sun's rays are directly overhead at noon on the Equator. (North of the **Arctic Circle**, at 66½ degrees North Latitude, there are 24 hours of daylight during the northern summer and 24 hours of darkness during the northern winter; south of the **Antarctic Circle**, at 66½ degrees South Latitude, there are 24 hours of daylight during the southern summer and 24 hours of darkness during the southern winter).

Remember that the parallels, or lines of latitude, make up only half the system used in the geographic grid; the meridians are necessary to complete the picture. **Meridians** are also called lines of longitude. A line of **longitude** is a straight line that circles the globe in a north-south direction at right angles to the parallels. The meridian that provides the point of reference for all other lines of longitudes is called the prime meridian (i.e., the primary or first meridian). The **prime meridian** is the line of longitude that runs directly from the North Pole to the South Pole through the Royal Astronomical Observatory in Greenwich, England. Unlike the Equator, the prime meridian is not determined by the natural conformation of the earth; instead, it is determined by arbitrary convention. (The location of the prime meridian was selected, of course, at a time when the British Empire was politically and militarily dominant in the world.) As you go north or south of the Equator, lines of longitude become progressively closer to each other until they converge at the North and South Poles; thus the distance between any two meridians changes with every change in latitude (unlike any two parallels, which are always a standard distance apart).

The prime meridian is designated the 0 degree of longitude, and all other lines of longitude are measured either east or west of the prime meridian. If you go due east or due west from any point on the prime meridian, you will go exactly one-half of the way around the globe before meeting someone who set out on the same line in the opposite direction. Therefore the maximum number of degrees used to designate any line of longitude is 180 (the 360 degrees of a circle divided by 2). Washington, D.C., for example, is located at 77 degrees West Longitude, while Canberra, Australia is situated at 149 degrees East Longitude. The prime meridian divides the world into two equal halves: the **Eastern Hemisphere**, the half of the earth that lies east of the prime meridian, and the **Western Hemisphere**, the half of the earth that lies west.

Halfway around the world from the prime meridian, at 180 degrees either East or West Longitude, lies the **180th Meridian**. Notice that the 180th Meridian runs through the eastern edge of the Fijian archipelago in the Southern Hemisphere. (Fiji is located north-northeast of North Island in New Zealand, at approximately 18 degrees south latitude; on the map on the following page, the Fijian archipelago appears as two small island shapes, the northernmost of which is touching the 180th Meridian.) For much of its length, the 180th Meridian is overlain by the **International Date Line**, which marks the point where each new day on earth begins at midnight. However, the International Date Line deviates in places from the true 180 degree meridian to avoid putting different parts of the same country in different dates; it jogs east through the Bering Strait, for example, to avoid separating northeastern Siberia from the rest of Russia, and then west around the Aleutian Islands to avoid dividing Alaska. At the exact moment of midnight on the International Date Line, there is only one date for the entire earth. After that moment, however, as the earth rotates to the east and the midnight meridian moves to the west, there are two calendar dates present on earth. The portion of the earth from the International Date Line westward to the current position of the midnight meridian falls under the newer of the two dates; the rest of the earth from the midnight meridian westward to the International Date Line falls under the older date.

The location of the North and South Poles can be specified simply by giving their latitudes, since all lines of longitude converge at the poles: 90 degrees North Latitude designates the North Pole, while 90 degrees South Latitude indicates the South Pole. To locate any other point on the earth's surface, however, it is necessary to indicate the intersection of a particular line of latitude with a particular line of longitude. By convention, the latitude is always specified first. Thus the location of Washington, D.C. is specified as 38° N 77° W, while the location of Canberra, Australia is described as 35° S 149° E. Further precision can be obtained by subdividing each degree of latitude or longitude into 60 minutes, and then subdividing each minute of latitude or longitude into 60 seconds.

There is one final point to keep in mind when it comes to the topic of mapping the earth: every flat map, regardless of the style of projection, necessarily involves some distortion of the true picture of the earth. (Globes are the only completely accurate representations of the surface of the earth, but globes are too cumbersome for most purposes.) Any effort to represent the three-dimensional spherical reality of the earth on a two-dimensional plane inevitably distorts both size and direction. On the map of the world that appears on the following page, for example, Greenland appears to be enormous, even though it is in fact smaller than the country of Sudan in northeastern Africa. The flat map distortion results from the fact that lines of longitude converge at the poles. As a result, the apparent distortion in true size becomes ever greater as you move north or south of the Equator. Flat maps distort direction as well. Looking at the map of the world on the following page, and keeping in mind that the shortest distance between two points is a straight line, it would appear that a direct flight from Miami, Florida to Paris, France would take off in an east-northeast direction over the Atlantic Ocean. In fact, however, the shortest flight path from Miami to Paris hugs the east coast of the United States all the way to Maine before crossing the Atlantic between Newfoundland and Ireland.

The map on the following page highlights the locations of the **Equator**, the **Tropics of Cancer** and **Capricorn**, the **Arctic** and **Antarctic Circles**, and the **Prime** and **180th Meridians**. Study the map closely while referring to your atlas. You should be able to locate these principal lines of latitude and longitude on a blank map of the world by orienting yourself to recognizable land forms. Notice that the Prime Meridian passes through the island of Great Britain, while the 180th Meridian cuts through northeastern Siberia (*not* through the Bering Strait). It's easy to remember that the Equator passes through Ecuador, since the name of this South American country is derived from the Spanish word for equator (it also passes just south of the small African country of Equatorial Guinea). Notice also that the Tropic of Cancer passes through the Straits of Florida separating the Florida Keys from Cuba, while the Tropic of Capricorn divides the southern fifth of Madagascar from the rest of the island. Finally, notice that the Arctic Circle passes immediately over the northernmost point of land in Iceland, while the Antarctic Circle bisects the Antarctic Peninsula and otherwise almost entirely encompasses the continent. Use your atlas to identify similar cues that will enable you to locate all of the world's most significant lines of latitude and longitude.

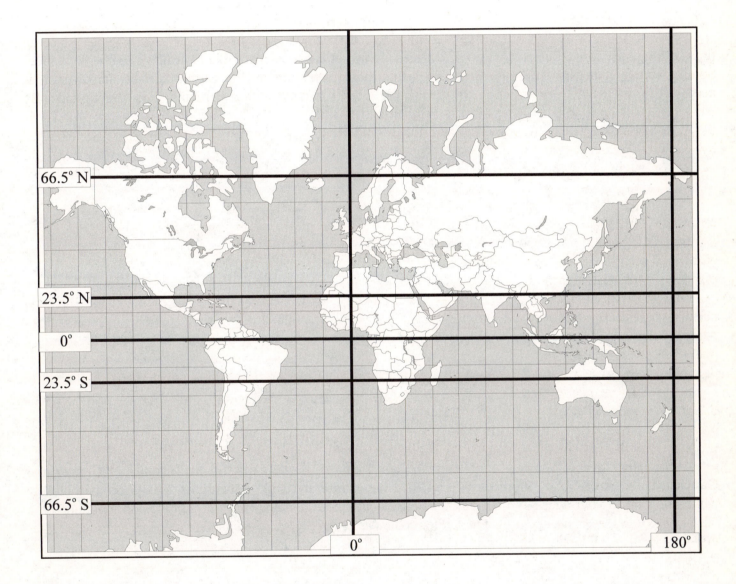

66.5° N

23.5° N

0°

23.5° S

66.5° S

0°   180°

Review Exercises
for
Planet Earth

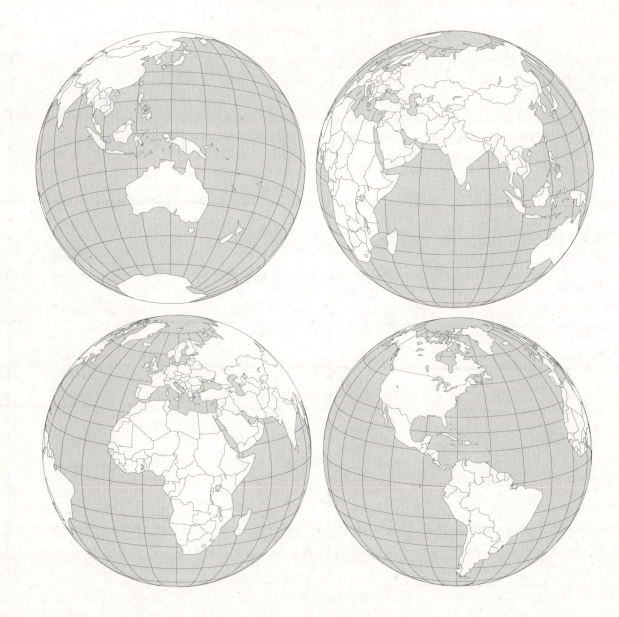

Solar System, Geographic Terms, Earth Dimensions, Landforms, Islands, Bodies of Water, Continents, Ten Largest Countries, Latitude & Longitude

1. List the eight planets of the solar system, in locational order beginning with the planet closest to the sun, and in size order, beginning with the largest planet:

Locational Order		Size Order	
1		1	
2		2	
3		3	
4		4	
5		5	
6		6	
7		7	
8		8	

2. A narrow strip of water that separates two landmasses and connects two larger bodies of water is called a(n) _____.

3. A wide strip of water that separates two landmasses and connects two larger bodies of water is called a(n)

    _____.

4. A narrow strip of land that separates two bodies of water and connects two larger landmasses is called a(n)

    _____.

5. A relatively large body of land that generally serves to define the borders of oceans is called a(n)

    _____.

6. A relatively small body of land that is entirely surrounded by water is called a(n)

    _____.

7. A point of land that extends into a body of water is called a(n)

    _____.

8. A piece of land that projects into a body of water and is surrounded by water on three sides is called a(n)

    _____.

9. A relatively small arm of a body of water that penetrates into a surrounding landmass is called a(n) _____.

10. An artificial channel or strait is called a(n) _____.

11. A relatively large arm of a body of water that penetrates into a surrounding landmass is called a(n) _____.

12. A body of water that is entirely surrounded by land and which has no natural outlet to the world ocean is called a(n) _____.

13. An extremely large body of water whose borders are defined by continental landmasses is called a(n) _____.

14. A relatively large portion of the ocean that is partially shaped by landmasses is called a(n) _____.

15. A collection of closely grouped islands is called a(n) _____.

16. The area at the mouth or lower course of a river where the river's current meets the sea's tide, mixing freshwater and saltwater, is called a(n) _____.

17. A long narrow arm of the sea bordered by steep cliffs, typically formed by glacial erosion of river valleys, is called a(n) _____.

18. The diameter of the earth is approximately _____ miles, and the circumference of the earth is approximately _____ miles.

19. List the world's seven continents in order of relative size, beginning with the continent with the largest geographic area. Make sure, of course, that you can locate each continent on a blank map of the world (additional blank maps can be found at the end of these *Review Exercises*).

1. _____
2. _____
3. _____
4. _____
5. _____
6. _____
7. _____

20.   Indicate the location of the following capes and peninsulas on the map of the world below.

Antarctic Peninsula   Cape Agulhas   Iberian Peninsula   Scandinavian Peninsula
Arabian Peninsula   Cape Horn   Italian Peninsula   Sinai Peninsula
Baja California   Cape of Good Hope   Korean Peninsula   Yucatan Peninsula
Balkan Peninsula   Florida Peninsula   Malay Peninsula

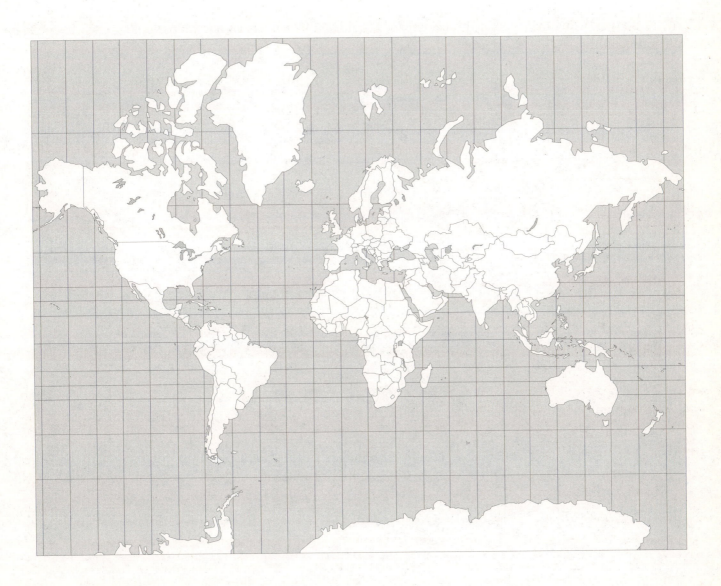

21.   Indicate the location of the following islands on the map of the world below.

Baffin	Great Britain	Java	New Guinea
Borneo	Greenland	Luzon	North Island
Celebes	Honshu	Madagascar	South Island
Cuba	Iceland	Mindanao	Sumatra
Ellesmere	Ireland	Newfoundland	Victoria

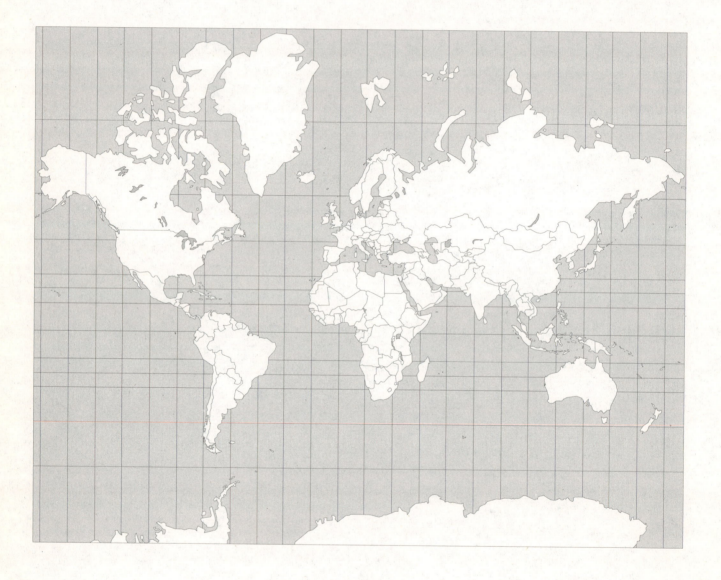

22.   Indicate the location of the following oceans and seas on the map of the world below.

Adriatic Sea	Bering Sea	Ionian Sea	Sea of Marmara
Aegean Sea	Black Sea	Mediterranean Sea	South China Sea
Arabian Sea	Caribbean Sea	North Sea	Southern Ocean
Arafura Sea	Caspian Sea	Norwegian Sea	Tasman Sea
Aral Sea	Coral Sea	Pacific Ocean	Timor Sea
Arctic Ocean	East China Sea	Red Sea	Tyrrhenian Sea
Atlantic Ocean	Indian Ocean	Sea of Japan	Yellow Sea
Baltic Sea			

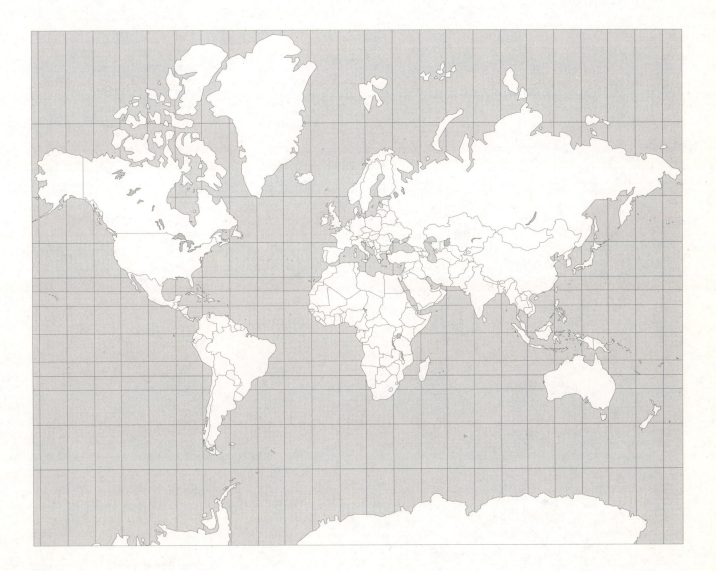

23. Indicate the location of the following bays, gulfs, canals, channels, lakes, and straits on the map of the world below.

Bay of Bengal	Gulf of Bothnia	Gulf of Tonkin	Lake Titicaca
Bay of Biscay	Gulf of California	Hudson Bay	Lake Victoria
Bering Strait	Gulf of Finland	Lake Baikal	Panama Canal
Bosporus	Gulf of Mexico	Lake Erie	Persian Gulf
Dardanelles	Gulf of Oman	Lake Huron	Strait of Gibraltar
Drake Passage	Gulf of St. Lawrence	Lake Michigan	Suez Canal
English Channel	Gulf of Suez	Lake Ontario	Strait of Hormuz
Gulf of Aden	Gulf of Thailand	Lake Superior	Strait of Malacca
Gulf of Aqaba			

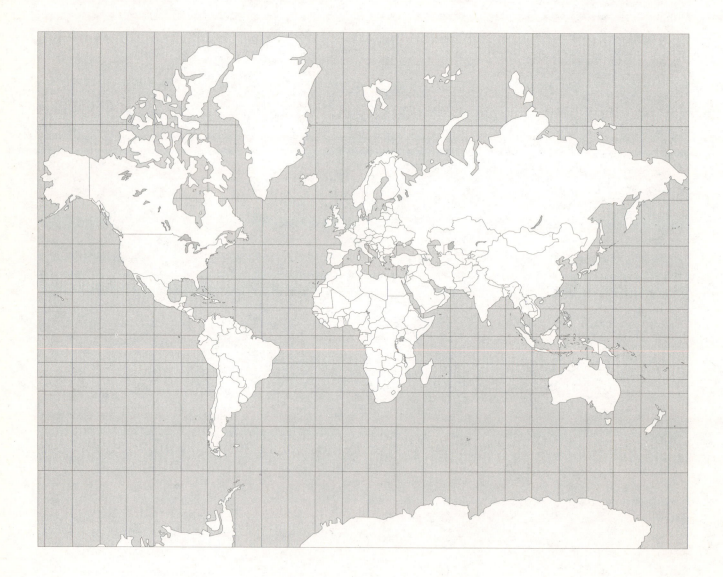

24.   List the ten largest countries in the world *by area*, and then indicate the location of each of the listed countries on the map of the world below.

Ten Largest Countries by AREA	
1	
2	
3	
4	
5	
6	
7	
8	
9	
10	

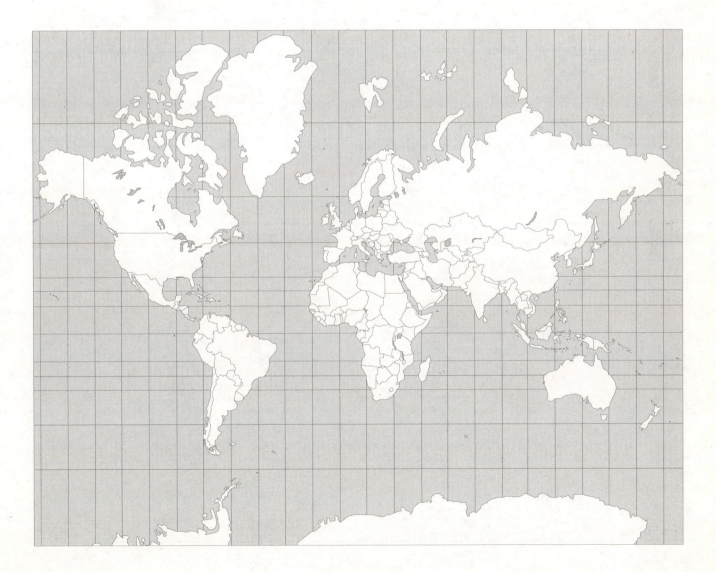

25. List the ten largest countries in the world *by population*, and then indicate the location of each of the listed countries on the map of the world below.

Ten Largest Countries by POPULATION	
1	
2	
3	
4	
5	
6	
7	
8	
9	
10	

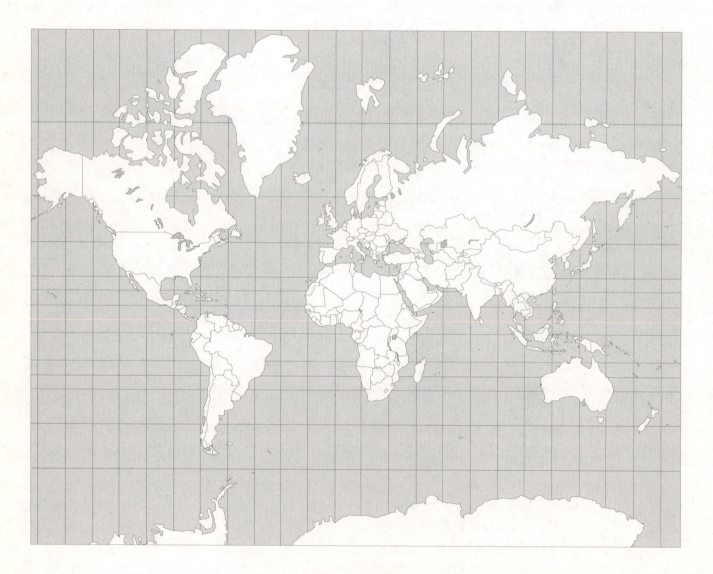

26.   On the map of the world below, indicate the location of the following principal lines of latitude and longitude (some lines will have to be drawn in using a ruler).

Latitude	Longitude
Equator Tropic of Cancer Tropic of Capricorn Arctic Circle Antarctic Circle	Prime Meridian 180th Meridian

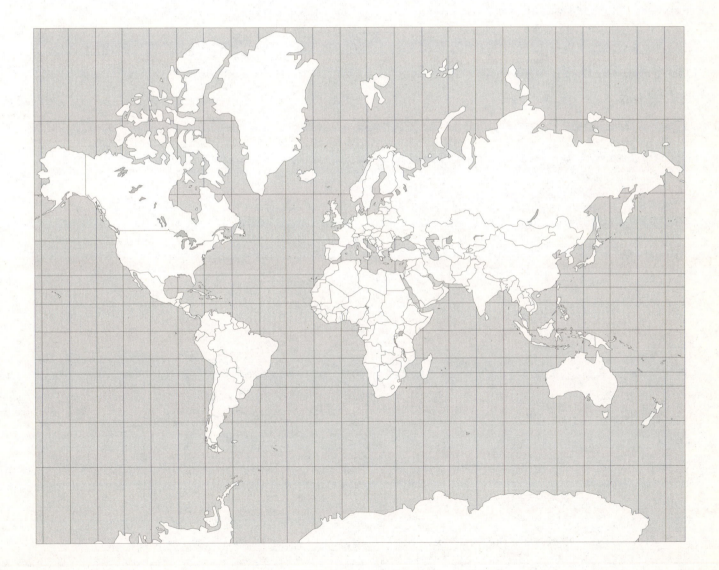

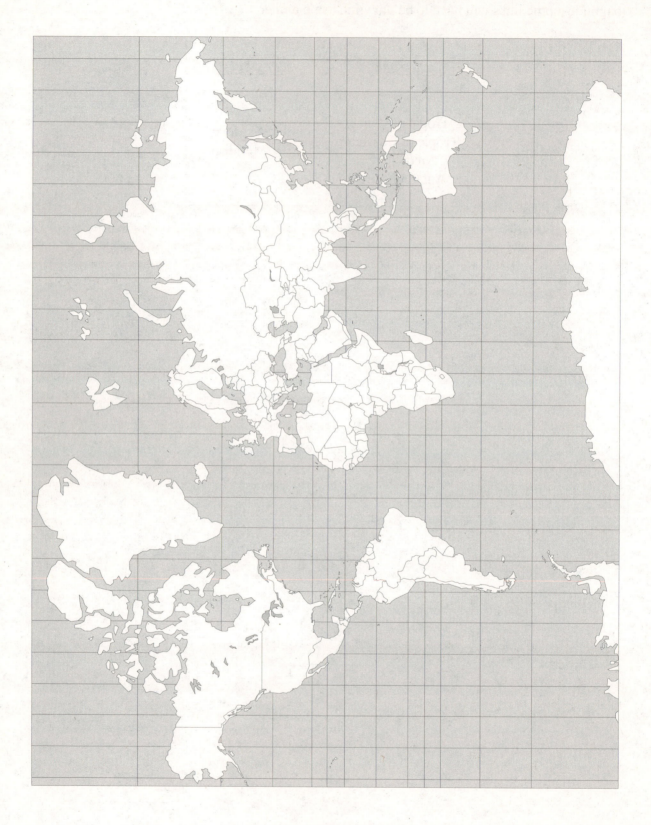

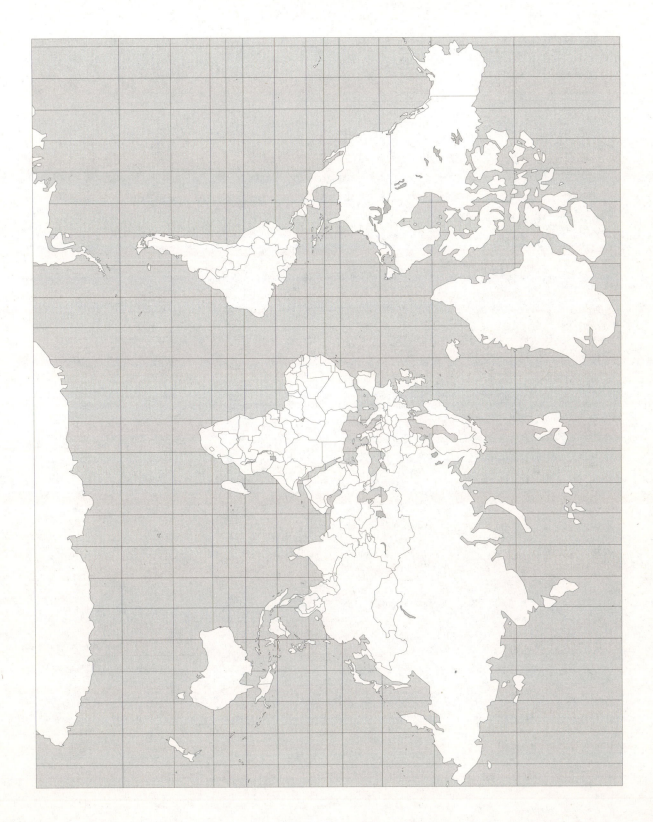

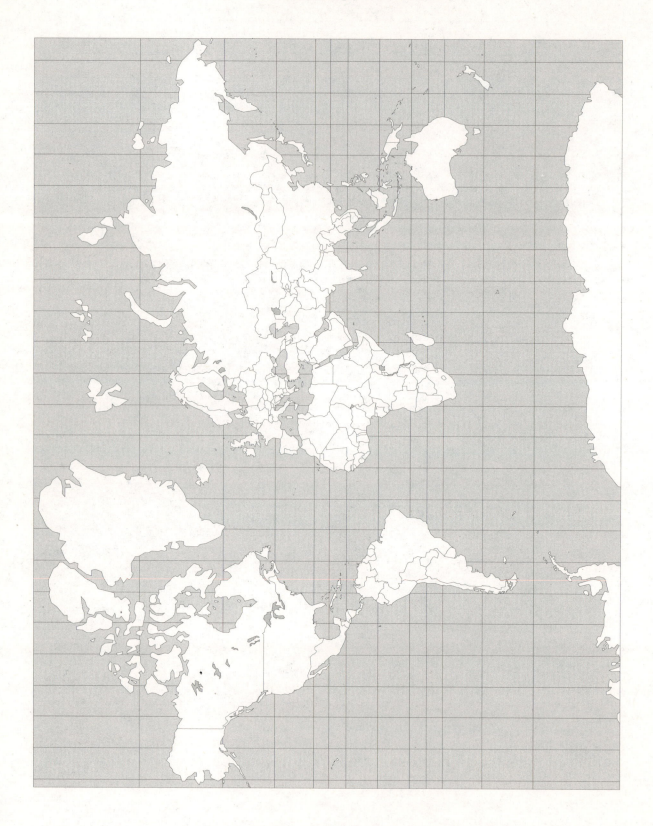

Places of the World

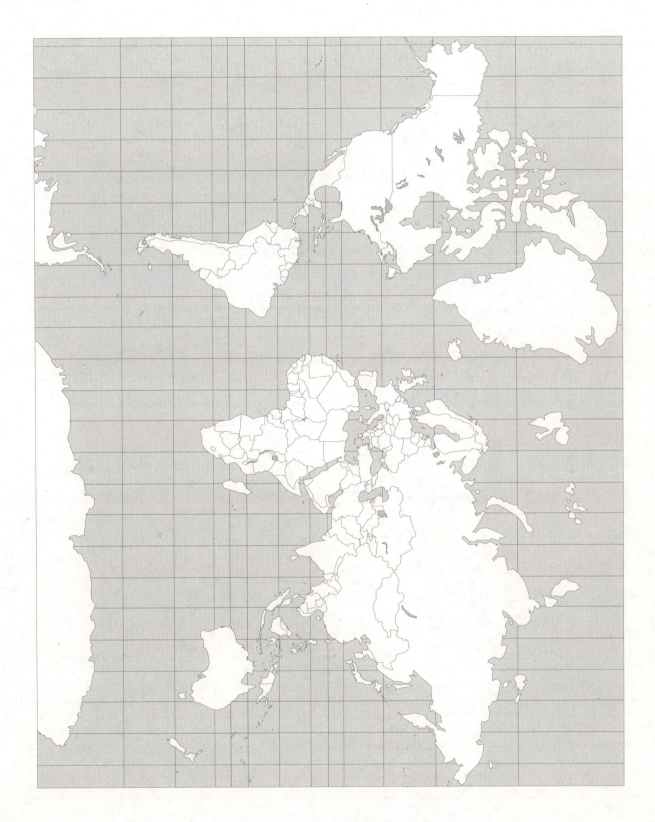

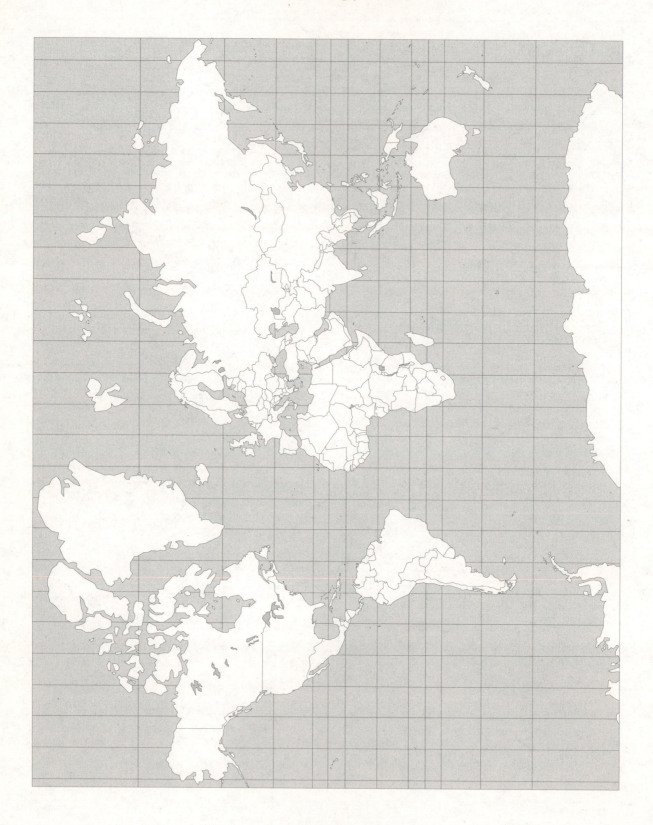

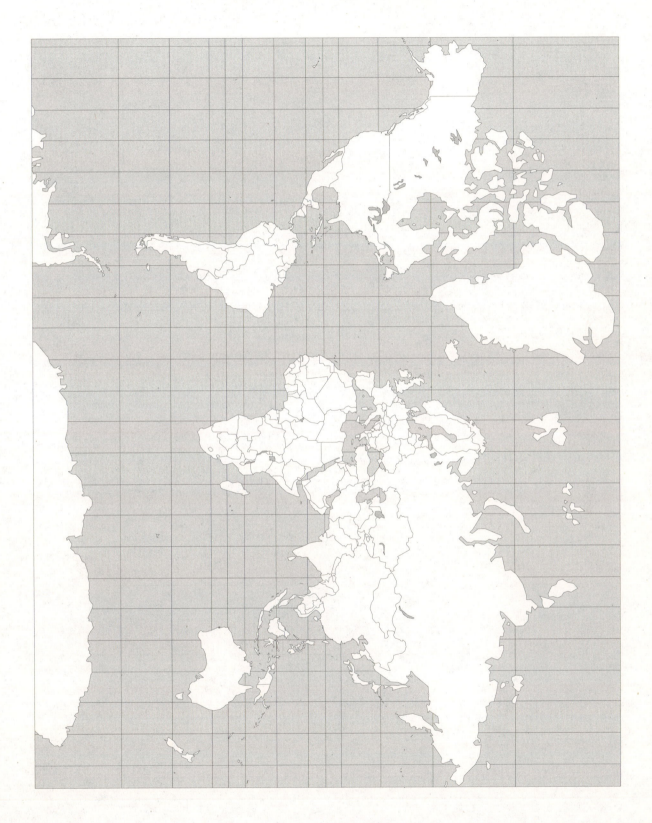

REGION 1
Europe

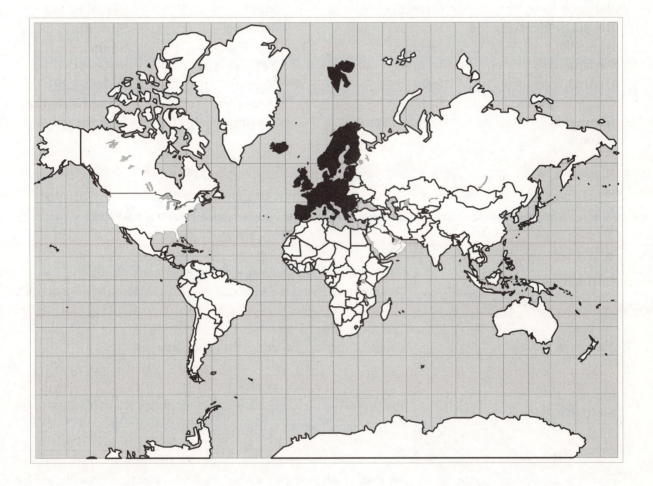

Europe

There are 41 independent countries in the European region covered in this chapter:

Albania	Finland	Lithuania	Romania
Andorra	France	Luxembourg	San Marino
Austria	Germany	Macedonia	Serbia
Belgium	Greece	Malta	Slovakia
Bosnia & Herzegovina	Hungary	Monaco	Slovenia
Bulgaria	Iceland	Montenegro	Spain
Croatia	Ireland	Netherlands	Sweden
Cyprus	Italy	Norway	Switzerland
Czech Republic	Latvia	Poland	United Kingdom
Denmark	Liechtenstein	Portugal	Vatican City
Estonia			

The 41 independent countries of Europe covered in this chapter can be conveniently divided into four subregions: the **European Core**, **Northern Europe**, **Southern Europe**, and **Eastern Europe**.

European Core	United Kingdom, Ireland, France, Germany, Belgium, Netherlands, Luxembourg, Switzerland, Austria, Liechtenstein, Monaco, Andorra
Northern Europe	Iceland, Norway, Sweden, Finland, Denmark
Southern Europe	Spain, Portugal, Italy, Malta, Greece, Cyprus, San Marino, Vatican City
Eastern Europe	Estonia, Latvia, Lithuania, Poland, Czech Republic, Slovakia, Hungary, Romania, Bulgaria, Slovenia, Croatia, Bosnia & Herzegovina, Serbia, Montenegro, Macedonia, Albania

Europe

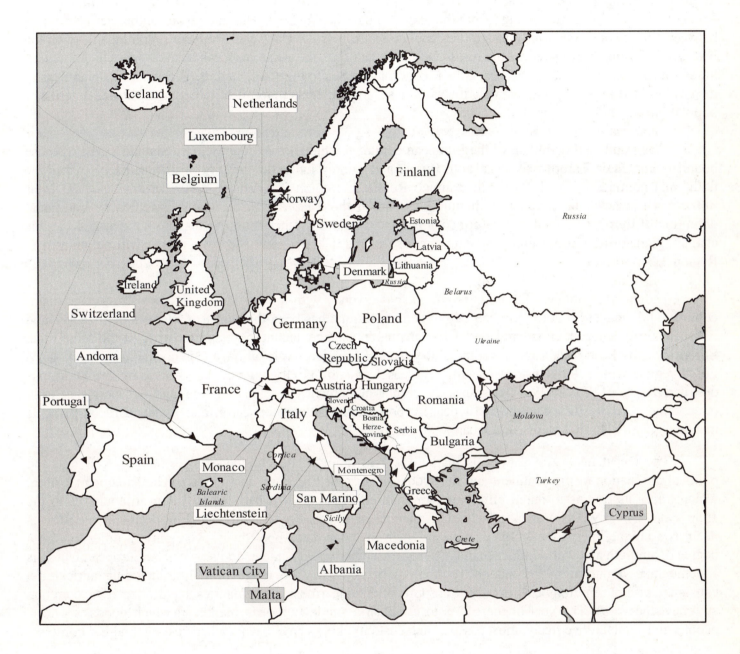

Iceland

Netherlands

Luxembourg

Belgium

Norway

Sweden

Finland

Estonia

Russia

Latvia

Lithuania

Denmark

Russia

Belarus

Ireland

United Kingdom

Germany

Poland

Switzerland

Czech Republic

Slovakia

Ukraine

Andorra

Austria

Hungary

Portugal

France

Italy

Slovenia

Croatia

Bosnia Herze- govina

Serbia

Romania

Moldova

Bulgaria

Spain

Corsica

Monaco

Balearic Islands

Sardinia

Montenegro

San Marino

Liechtenstein

Sicily

Greece

Turkey

Cyprus

Crete

Macedonia

Vatican City

Albania

Malta

Europe comprises the western portion of the vast Eurasian land mass and consists primarily of a series of peninsulas and associated islands in the surrounding waters. The continent extends north to south from beyond the Arctic Circle in Scandinavia to the Greek island of Crete in the southern Mediterranean. Europe is traditionally defined as extending west to east from Iceland in the North Atlantic to the Ural Mountains in Russia (the Urals run north-to-south from the Arctic to the Caspian Sea). The traditional southeastern boundary between Europe and Asia runs through the Caspian Sea, across the Caucasus Mountains, and then through the Black Sea. Europe is bordered by the Arctic Ocean to the north, the Atlantic Ocean to the west (which includes the Norwegian, North, and Baltic Seas), the Mediterranean Sea to the south, and the Black and Caspian Seas to the east. Europe ranks 6th in size among the world's seven continents in terms of area, and 3rd in size in terms of population.

This traditional definition of the European continent includes two countries, Russia and Turkey, whose boundaries straddle Europe and Asia (both Russia and Turkey have the majority of their territory within the traditional confines of Asia). We will examine Russia in Region 2 when we review *Russia and the Near Abroad*; Turkey will be covered in Region 3 as part of the *Middle East and North Africa*. The traditional definition of European boundaries described above includes a few other countries that are not discussed in this chapter, namely the Eastern European countries of Belarus, Ukraine, and Moldova, which will be covered in Region 2.

The small continent of Europe is remarkably diverse, with a wide variety of climates, topographies, cultures, and languages. For more than two millennia, Europe has been the center of Western civilization, and Europeans can justly claim unparalleled achievements in science, technology, and the arts. At the same time, Europeans have been responsible for some of the greatest injustice, exploitation, and genocide in human history. For a region with a long history of virtually ceaseless conflict (highlighted, of course, by the enormous destructiveness of World War II), Europe's recent attainment of relative peace and prosperity is truly remarkable. Today, Europe is the most densely populated continent on earth, and its standard of living is matched only by North America and Japan.

The recent history of Europe is marked by two dramatic developments: the ongoing economic and political unification of the continent and the sudden demise of European communism. The **European Union**, which traces its origins to initial efforts at economic cooperation by a handful of European countries in the 1950's, had by early 2008 grown to include 27 member states (see *Appendix B*). The European Union describes itself as a unique international organization in which member states pool their sovereignty in order to gain a degree of strength and world influence that no single member could achieve on its own. The European Union is a democratic organization that adopts common policies in a wide range of fields, including agriculture, consumer affairs, environment, energy, transport, and trade (most EU countries have adopted a common currency, the *euro*). The goal of the EU is to establish a single European market in which goods, services, people, and capital move freely; most passport and customs checks have been abolished at internal EU borders.

Since the destruction of the Berlin Wall in 1989 and the collapse of the Soviet Union in 1991, the major story in eastern Europe has been the replacement of authoritarian socialist regimes by democratic capitalist governments. The story is a checkered one, however, marred most notably by the vicious civil war in the former Yugoslavia. Recent electoral and economic developments elsewhere in Europe suggest that the eventual emergence of a fully democratic and fully capitalistic Europe is still somewhat problematic.

European Core

The twelve countries of the European Core can be further subdivided into the **British Isles**, **France & Germany**, **Benelux**, and the **Alpine States**, plus three of Europe's **Microstates**.

British Isles

Scotland

Northern
Ireland

Ireland

United
Kingdom

England

Wales

The British Isles consist of two major landforms, Great Britain and Ireland, and a large number of small surrounding islands. Most of the island of Ireland is ruled by the predominantly Roman Catholic Republic of Ireland. The country's economy was for a long time more agrarian and less developed than most of the rest of northwestern Europe, and as a result Ireland experienced high rates of emigration throughout much of the 20th century. In recent years, however, Ireland's economy has been growing rapidly, fueled by a thriving export industry. Northern Ireland, which is predominantly Protestant, is part of the United Kingdom, the country that includes the entire island of Great Britain as well. (Great Britain itself is further divided into the historical units of England, Wales, and Scotland.) The narrow English Channel separating Great Britain from the European mainland can now be traversed by rail via the Channel Tunnel, colloquially known as the "Chunnel."

UNITED KINGDOM

Population	60,776,000 *(648 per square mile)*
Language(s)	English
Area	93,788 *square miles*
Capital	**London**
Economy	$31,800 *annual per capita GDP*
Sovereignty	Constitutional Monarchy, *independent national identity since* 1066

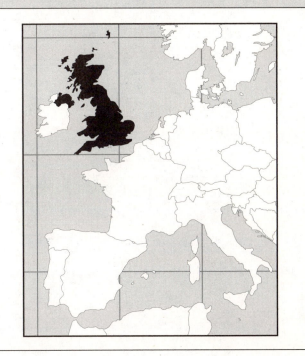

IRELAND

Population	4,109,000 *(151 per square mile)*
Language(s)	English & Gaelic
Area	27,133 *square miles*
Capital	**Dublin**
Economy	$44,500 *annual per capita GDP*
Sovereignty	Democratic Republic *independence established* 1921

France & Germany

Corsica

The two large countries of France and Germany lie in the historic and geographic heart of mainland western Europe. With a population of more than 80 million people, Germany is the most populous country located entirely within the boundaries of the European continent (excluding Russia, whose territory overlaps the traditional boundary between Europe and Asia). With just under 61 million people, France has the second largest population in Europe (the United Kingdom and Italy are close behind). France is also the second largest country entirely within the traditional borders of Europe in terms of geographic size, ranking just behind Ukraine in land area. Germany and France are both highly developed centers of manufacturing, commerce, and finance (at the present time, Germany has the most productive economy in Europe). Notice that the Mediterranean island of Corsica is part of France.

FRANCE

Population	60,876,000 *(288 per square mile)*
Language(s)	French
Area	208,482 *square miles*
Capital	**Paris**
Economy	$31,200 *annual per capita GDP*
Sovereignty	Democratic Republic, *independent since unification by Clovis in 486*

GERMANY

Population	82,401,000 *(598 per square mile)*
Language(s)	German
Area	137,847 *square miles*
Capital	**Berlin**
Economy	$31,900 *annual per capita GDP*
Sovereignty	Federal Democratic Republic independence established 1871

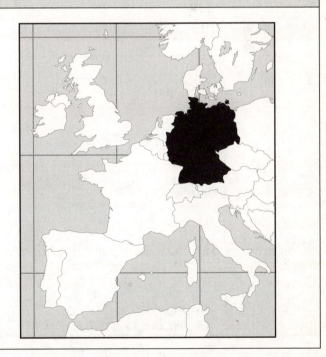

Benelux

The collective term Benelux is derived from the first syllables of the three countries to which it refers: *Bel*gium, *Net*herlands, and *Lux*embourg. These three states are also called the Low Countries, because they are comprised primarily of flat land that lies very close to sea level (and even below sea level in some parts of the Netherlands, where the land has been reclaimed from the North Sea and lies protected behind extensive systems of dikes). The population density, especially in the Netherlands and Belgium, is the highest in Europe outside the Microstates.

The history of Belgium has been largely determined by its geographic location. This small country lies at the crossroads of European powers, and has long served as a battlefield for European wars. Napoleon met his Waterloo in Belgium. The Netherlands is frequently referred to as Holland, although the term is technically incorrect, because it confuses the part for the whole. The Netherlands is divided into eleven provinces, two of which are named North Holland and South Holland. Referring to the entire country as Holland, then, would be comparable to calling Canada "Ontario." Amsterdam, the official capital city of the Netherlands, is the home of the reigning monarch, Queen Beatrix, but the judicial and parliamentary seat of government is located in The Hague. The tiny country of Luxembourg has achieved a degree of economic and political influence that is disproportionate to its size. An important international banking center, Luxembourg enjoys the highest per capita GDP in the world, and the country was an influential charter member of the multinational organization that eventually became the European Union. Although diminutive in both geographic area and population, Luxembourg is far from the smallest country in Europe; it is substantially larger, in fact, than all five of the European Microstates.

BELGIUM

Population	10,392,000 *(882 per square mile)*	
Language(s)	Dutch, French, & German	
Area	11,787 *square miles*	
Capital	**Brussels**	
Economy	$33,000 *annual per capita GDP*	
Sovereignty	Constitutional Monarchy *independence established* 1830	

NETHERLANDS

Population	16,571,000 *(1,025 per square mile)*
Language(s)	Dutch and Frisian
Area	16,164 *square miles*
Capital	**Amsterdam** (capital) and **The Hague** (seat of government)
Economy	$32,100 *annual per capita GDP*
Sovereignty	Constitutional Monarchy *independence established* 1648

LUXEMBOURG

Population	480,000 *(481 per square mile)*
Language(s)	Letzeburgesch, German, & French
Area	999 *square miles*
Capital	**Luxembourg** (*sometimes referred to as* Luxembourg-Ville)
Economy	$71,400 *annual per capita GDP*
Sovereignty	Constitutional Monarchy *independence established* 1839

Alpine States

   The landlocked states of Switzerland and Austria are both dominated by the scenic splendor of the highest mountains in Europe, the Alps, but otherwise the two countries have relatively little in common. Switzerland is a multi-lingual country with a long history of political stability and independence. The Swiss economy is one of the most productive in the world on a per capita basis, and the Swiss enjoy one of the world's highest standards of living. Austria, on the other hand, is a unilingual country with a long history of political turmoil and foreign occupation, and an economy that is relatively inefficient compared to Switzerland's.

SWITZERLAND

Population	7,555,000 *(474 per square mile)*
Language(s)	German, French, & Italian
Area	15,943 *square miles*
Capital	**Bern**
Economy	$34,000 *annual per capita GDP*
Sovereignty	Federal Democratic Republic *independence established* 1291

AUSTRIA

Population	8,200,000 *(253 per square mile)*
Language(s)	German
Area	32,378 *square miles*
Capital	**Vienna**
Economy	$34,700 *annual per capita GDP*
Sovereignty	Federal Democratic Republic *independence established* 1156

Microstates

The five "microstates" of Europe (Andorra, Liechtenstein, San Marino, Monaco, and Vatican City) include some of the smallest and oldest independent countries in the world. Although all five are universally recognized as fully sovereign states, all five have maintained close political and economic associations with their larger European neighbors. The sovereignty of Andorra, for example, is recognized by the encircling countries of France and Spain, but Andorra in turn officially recognizes the French President and the Spanish Bishop of Urgel as its joint heads of state. Liechtenstein followed the example of its cultural counterpart Switzerland and remained neutral during World War II. As an international financial center, Liechtenstein has emulated Switzerland economically as well, and enjoys a relatively high per capita GDP. The people of San Marino are ethnically and culturally Italian, although they have a strong sense of national identity, derived in part from the fact that theirs is the oldest country in Europe. The principality of Monaco owes its independence to treaties with France, just as Vatican City owes its independence to treaties with Italy.

Liechtenstein, Monaco, and Andorra are described in this section on the European Core; San Marino and Vatican City will be described in the section on Southern Europe.

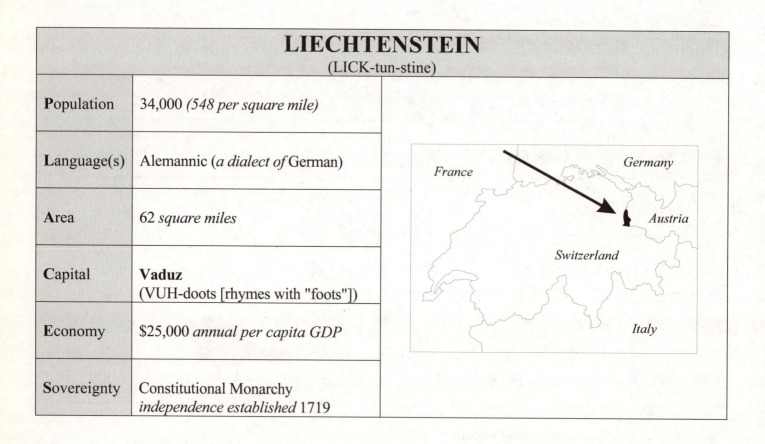

LIECHTENSTEIN
(LICK-tun-stine)

Population	34,000 *(548 per square mile)*	
Language(s)	Alemannic *(a dialect of* German)	
Area	62 *square miles*	
Capital	**Vaduz** (VUH-doots [rhymes with "foots"])	
Economy	$25,000 *annual per capita GDP*	
Sovereignty	Constitutional Monarchy *independence established* 1719	

MONACO

Population	33,000 *(44,000 per square mile)*
Language(s)	French, English, Italian, and Monegasque
Area	.75 *square miles*
Capital	**Monaco** (*a.k.a.* Monaco-ville)
Economy	$30,000 *annual per capita GDP*
Sovereignty	Constitutional Monarchy *independence established* 1419

ANDORRA

Population	72,000 *(398 per square mile)*
Language(s)	Catalan, Spanish, & French
Area	181 *square miles*
Capital	**Andorra la Vella**
Economy	$38,800 *annual per capita GDP*
Sovereignty	Parliamentary Democracy *independence established* 1278

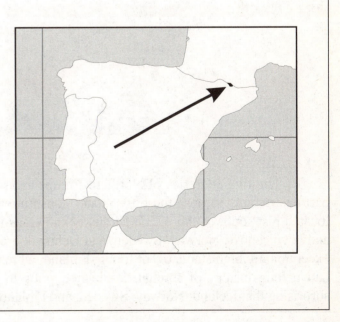

Northern Europe

Iceland

Sweden

Finland

Norway

Denmark

The five countries of Northern Europe are sometimes referred to as the "Nordic States." The name is particularly apt, because these five states constitute the northernmost group of countries in the world. Iceland, located just south of the Arctic Circle, is known as the "land of ice and fire" because of its frigid climate and active volcanoes. Norway, Sweden, and Denmark are collectively referred to as Scandinavia, and all three enjoy exceptionally high standards of living. Finland, which declared its independence from the Soviet Union in 1917, had a long history of foreign domination prior to the 20th century. Given their extreme latitude, it is not surprising that Iceland, Norway, Sweden, and Finland have the lowest population densities in Europe.

ICELAND

Population	302,000 *(8 per square mile)*
Language(s)	Icelandic
Area	39,769 *square miles*
Capital	**Reykjavik** (RAY-kyuh-vihk)
Economy	$38,000 *annual per capita GDP*
Sovereignty	Democratic Republic *independence established* 1944

NORWAY

Population	4,628,000 *(37 per square mile)*
Language(s)	Norwegian
Area	125,050 *square miles*
Capital	**Oslo**
Economy	$46,300 *annual per capita GDP*
Sovereignty	Constitutional Monarchy *independence established* 1905

SWEDEN

Population	9,031,000 *(52 per square mile)*
Language(s)	Swedish
Area	173,732 *square miles*
Capital	**Stockholm**
Economy	$32,200 *annual per capita GDP*
Sovereignty	Constitutional Monarchy *independence established* 1523

FINLAND

Population	5,238,000 *(40 per square mile)*
Language(s)	Finnish & Swedish
Area	130,559 *square miles*
Capital	**Helsinki**
Economy	$33,500 *annual per capita GDP*
Sovereignty	Democratic Republic *independence established* 1917

DENMARK

Population	5,468,000 *(329 per square mile)*
Language(s)	Danish
Area	16,640 *square miles*
Capital	**Copenhagen**
Economy	$37,100 *annual per capita GDP*
Sovereignty	Constitutional Monarchy *independence established* 1849

Southern Europe

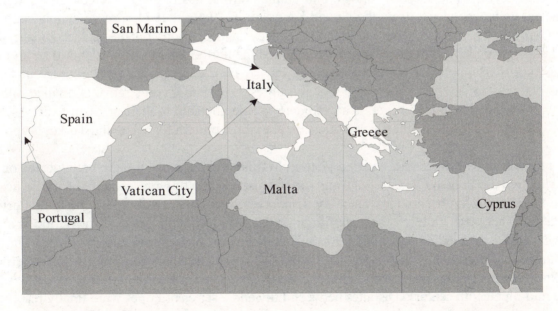

The eight countries of Southern Europe can be further subdivided into the **Iberian Peninsula,** the **Italian Peninsula**, and the **Central & Eastern Mediterranean**, plus the remaining two of Europe's **Microstates**.

The **Iberian Peninsula** is geographically separated from the rest of Europe by the Pyrenees Mountains, and the lightly industrialized economies of Spain and Portugal have traditionally been isolated from European prosperity as well. Since the mid 1980's, however, Spain's economy has exhibited dramatic growth, particularly in the tourism industry, and the Mediterranean coast of Spain is becoming an increasingly popular retirement site for people throughout northwestern Europe. In climate, topography, and culture, Spain and Portugal have more in common with the Mediterranean cultures of southern Italy and Greece than they do with the societies of northwestern Europe. Notice that the Balearic Islands of Majorca, Minorca, and Ibiza, located in the Mediterranean Sea off the Spanish coast, are an integral part of Spain.

With the exception of the two tiny enclave countries of San Marino and Vatican City, the entire **Italian Peninsula** is occupied by the country of Italy. The northern half of the country has a highly developed economy comparable to that of France and Germany; the southern half is less developed and more agrarian. Notice that the Mediterranean islands of Sardinia and Sicily are also part of Italy.

Two of the countries of Southern Europe, Greece and Malta, can be found in the **Central Mediterranean**. Greece occupies the southern tip of the Balkan Peninsula, but the country includes as well more than 3,500 islands in the surrounding Aegean and Ionian Seas. Greece's economy has traditionally been based on agriculture, shipping, and tourism, but increasing industrialization during the past three decades has been accompanied by one of the highest rates of inflation in the European Union. The tiny island country of Malta was a British colony for more than 150 years before becoming independent in 1964; since declaring itself a republic in 1974, however, it has retained its membership in the Commonwealth (see *Appendix B*).

The island nation of Cyprus, located in the **Eastern Mediterranean**, is a divided country. About $4/5^{th}$ of the population is Greek while $1/5^{th}$ is Turkish, and ethnic conflict has reigned between the two groups since Cyprus achieved its independence from the United Kingdom in 1960. In 1974, Greece announced its desire to incorporate Cyprus as part of Greece; in response, Turkey landed troops on the island, and heavy fighting followed. The following year, Turkish Cypriots proclaimed the northern part of the island to be an independent country (the Turkish Republic of Northern Cyprus), but so far Turkey is the only country in the world to recognize its claim to sovereignty. The conflict is still unresolved, and Cyprus remains divided into a Turkish zone and a Greek zone. Turkey, however, is a candidate for membership in the European Union (see *Appendix B*), and the EU has made resolving the Cyprus question one of the criteria for accepting Turkey's application.

The remaining two **Microstates** of Europe (San Marino and Vatican City) are both located on the Italian Peninsula. Although the French Chief of State and the Spanish Bishop of Urgel remain the titular heads of state for Andorra, the country has been a parliamentary democracy since 1993; with a thriving tourist economy and no income taxes, Andorra is a mecca for many immigrants, both legal and illegal. San Marino is arguably the world's oldest republic; according to tradition, it was founded by a Christian stonemason named Marinus in 301 C.E. San Marino's foreign policy is closely tied to Italy's. In their secular role, Roman Catholic Popes ruled portions of Italy for over a thousand years, until the newly united Kingdom of Italy seized many of the Papal States in the mid-nineteenth century. Disputes between Italy and a series of Popes were finally resolved in 1929, when the Lateran Treaties established the independent state of Vatican City.

SPAIN

Population	40,448,000 *(208 per square mile)*
Language(s)	Spanish
Area	194,885 *square miles*
Capital	**Madrid**
Economy	$27,400 *annual per capita GDP*
Sovereignty	Constitutional Monarchy, *independent national identity since* 1492

PORTUGAL

Population	10,643,000 *(300 per square mile)*
Language(s)	Portuguese
Area	35,516 *square miles*
Capital	**Lisbon**
Economy	$19,800 *annual per capita GDP*
Sovereignty	Democratic Republic *independence established* 1910

ITALY

Population	58,148,000 *(500 per square mile)*
Language(s)	Italian
Area	116,342 *square miles*
Capital	**Rome**
Economy	$30,200 *annual per capita GDP*
Sovereignty	Democratic Republic *independence established* 1861

MALTA

Population	402,000 *(3,295 per square mile)*
Language(s)	Maltese & English
Area	122 *square miles*
Capital	**Valletta** (vahl-LEHT-uh)
Economy	$21,300 *annual per capita GDP*
Sovereignty	Democratic Republic *independence established* 1964

GREECE

Population	10,706,000 *(210 per square mile)*
Language(s)	Greek
Area	50,949 *square miles*
Capital	**Athens**
Economy	$24,000 *annual per capita GDP*
Sovereignty	Democratic Republic *independence established* 1829

CYPRUS
(SIGH-pruss)

Population	788,000 *(221 per square mile)*
Language(s)	Greek & Turkish
Area	3,572 *square miles*
Capital	**Nicosia** (nik-oh-SEE-uh) *[the Turkish-occupied portion of Nicosia is called* Lefkosa]
Economy	$23,000 *annual per capita GDP*
Sovereignty	Republic *independent since* 1960

SAN MARINO

Population	30,000 *(1,250 per square mile)*
Language(s)	Italian
Area	24 *square miles*
Capital	**San Marino**
Economy	$34,100 *annual per capita GDP*
Sovereignty	Republic, *independent since* 301

VATICAN CITY
(*also known as* The Holy See)

Population	820 *(4,824 per square mile)*
Language(s)	Latin & Italian
Area	.17 *square miles*
Capital	**Vatican City**
Economy	*N/A per capita GDP.* This unique, noncommercial economy is supported by contributions from Roman Catholics throughout the world, the sale of postage stamps and tourist mementos, fees for admission to museums, and the sale of publications. The Holy See also has worldwide banking and financial activities that generate substantial revenue.
Sovereignty	Ecclesiastical State, *independent since* the 1929 Lateran Treaties with Italy

Eastern Europe

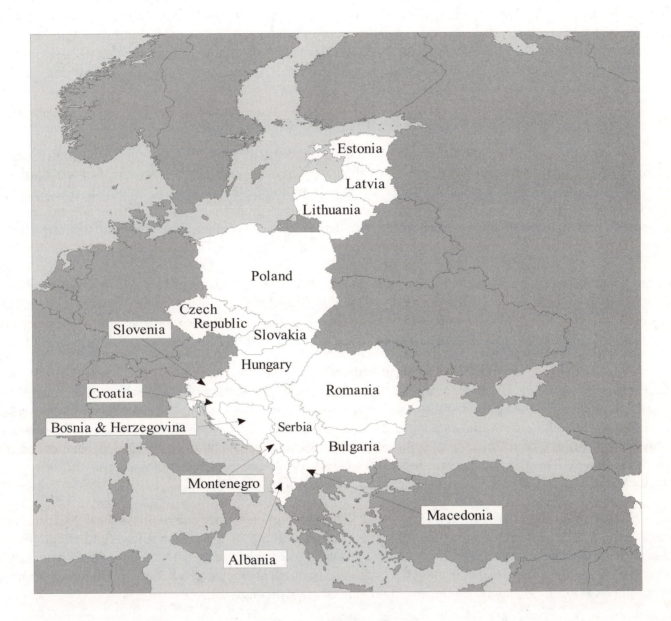

The three **Baltic States** of Estonia, Latvia, and Lithuania can be found in the northernmost portion of Eastern Europe, while the four countries of **Poland, Czech Republic, Slovakia, and Hungary** lie in the heart of the region in the area that forms the border between the European Core and Russia & the Near Abroad; the remaining nine countries of Eastern Europe occupy the **Balkan Peninsula** on the southeastern extremity of Europe.

Baltic States

Although culturally and linguistically distinct from each other, the three small "Baltic States" of Estonia, Latvia, and Lithuania have considerable history in common. All three were occupied by various powers prior to the 20th century (including Germany, Poland, Sweden, and Russia); all three achieved independence following World War I; all three were forcibly absorbed into the Soviet Union in 1940; and all three reacquired their independence in 1991. Today, Estonians, Latvians, and Lithuanians make up the majority ethnic groups in their respective states, but all three countries are ethnically diverse: Russians, Ukrainians, Belorussians, and Poles are well represented among the populations of the three small countries that border the Baltic Sea. (Russia and the three other former Soviet republics in Eastern Europe—Belarus, Ukraine, and Moldova—will be described in Region 2 dealing with *Russia and the Near Abroad*.)

Poland, Czech Republic, Slovakia, and Hungary

The present geographic boundaries of Poland date to the end of World War II, when the country gained territory on its northern and western borders at the expense of Germany and lost territory on its eastern border to the former Soviet Union. With a population approaching 40 million people, Poland is larger than three-quarters of the countries in Europe. In comparison with most other former communist states, Poland has made more rapid strides in the transformation to capitalism, but large numbers of Polish citizens still face significant economic hardship (unemployment has been high in recent years, and the GDP has been growing slowly). The impact of growing economic difficulties were dramatically reflected in the December 1995 Polish presidential election, when Lech Walesa, who had helped topple the communist government in 1990, was replaced by a former communist official; Walesa's Solidarity Trade Union suffered a devastating setback in the 2001 Parliamentary elections when it failed to elect a single deputy to the lower house of Parliament. Poland became a member of NATO in 1999 and a member of the European Union in 2004 (see *Appendix B*).

The dissolution of the former independent country of Czechoslovakia in 1993 produced two of Europe's newest sovereign states, the Czech Republic and Slovakia. The fragmentation of Czechoslovakia along cultural and linguistic lines paralleled events in the former countries of Yugoslavia and the Soviet Union and may serve as a harbinger for political developments in the 21st century, except that the dissolution of Czechoslovakia was notably peaceful (the Czech Republic and Slovakia owe their separate identities to what has been called the "Velvet Divorce"). The Czech Republic joined NATO in 1999, and Slovakia followed suit in 2004; both countries became members of the European Union in 2004 (see *Appendix B*).

Hungary enjoys a relatively high standard of living among the former communist states of Central and Eastern Europe. The country is agriculturally self-sufficient and one of the region's few food exporters. Its capital city of Budapest, strategically located on the Danube River, is several times larger than Hungary's second most populous urban area, indicating the traditional agrarian character of the country. Hungary became a member of NATO in 1999 and the European Union in 2004 (see *Appendix B*).

The Balkan Peninsula

Romania, Bulgaria, Slovenia, Croatia, Bosnia & Herzegovina, Serbia, Montenegro, Macedonia, and Albania all occupy the Balkan Peninsula—the large triangular landmass whose three points are the northeastern corner of the Adriatic Sea, the northwestern corner of the Black Sea, and the southern tip of the Greek Peloponnesus. For centuries, the region has been home to a wide diversity of peoples and cultures that have frequently found themselves in conflict with one another and with foreign invaders. The recent violent dismantling of the former Yugoslavia is simply the latest installment in a long-standing story: the history of the

Balkan Peninsula is a history of conflict, fragmentation, and shifting national boundaries. As a result, the term "balkanization" has entered the English language to refer to the division of larger entities into smaller, quarrelsome, and ineffectual units.

Prior to the collapse of communism in Europe, which began in 1989, the multi-ethnic country of Yugoslavia was comprised of six constituent republics: Slovenia, Croatia, Bosnia-Herzegovina, Macedonia, Serbia, and Montenegro. By the early 1990's, the former socialist republic of Yugoslavia had fragmented into five independent states (Serbia and Montenegro remained united as the Federal Republic of Yugoslavia), but the political boundaries that emerged encompassed a mosaic of ethnic, religious, and linguistic differences. In December 1995, after five years of vicious civil war that marked the introduction of the phrase "ethnic cleansing" into the English language, the Slovenes, Croats, Bosnians, and Serbs finally signed a putative peace agreement, but a brutal Serbian campaign against ethnic Albanians in the Yugoslav province of Kosovo led to NATO air attacks against Yugoslavia beginning in March 1999. In October 2000, Yugoslav President Slobodan Milosevic was ousted following blatant attempts to manipulate presidential balloting that resulted in massive nationwide demonstrations; Milosevic was subsequently brought before an International Criminal Tribunal in The Hague to stand trial on charges of war crimes (he died in 2006 before the completion of his trial). In 2002, Yugoslavia changed its name to *Serbia and Montenegro*, and then in 2006 Montenegro and Serbia declared their independence from each other, forming two more separate countries from the final remnant of the former Yugoslavia. By early 2008, the Serbian province of Kosovo was moving rapidly toward independence.

Croatia has been relatively peaceful since 1998, when the last Serb-held enclave in the country was returned to Croatian control. Macedonia's large Albanian minority staged an armed insurgency in 2001 which lead to an internationally-brokered Framework Agreement, but that agreement has yet to be fully implemented. In Bosnia & Herzegovina , European Union peacekeeping forces replaced a NATO-led Stabilization Force in 2004, and began winding down their mission in 2007. Of the former Yugoslav countries, Slovenia has so far made the greatest progress towards becoming a peaceful, prosperous member of the world community—it joined NATO and the European Union in 2004 (Croatia's EU membership application is pending).

The former communist country of Albania has long had a reputation as one of the least developed countries in Europe. It is a small, mountainous country with relatively few natural resources, high unemployment, and pervasive poverty. Albania is predominantly Muslim, and it has an especially high population growth rate compared to other European countries. Albania faces ongoing problems stemming from widespread corruption, a dilapidated infrastructure, and powerful organized crime networks with links to high government officials.

The northeastern portion of the Balkan Peninsula is occupied by the countries of Bulgaria and Romania. Both countries border the western shore of the Black Sea, but otherwise they have little in common (although both are former communist countries, and both have relatively low standards of living). Bulgarians speak a Slavic language, Romanians a Romance language, and the country of Bulgaria is less than half the size of its northern neighbor both geographically and demographically.

ESTONIA
(ess-TOH-nee-uh)

Population	1,316,000 *(75 per square mile)*
Language(s)	Estonian
Area	17,462 *square miles*
Capital	**Tallinn** (TAHL-inn)
Economy	$20,300 *annual per capita GDP*
Sovereignty	Democratic Republic, *independence reestablished* 1991

LATVIA

Population	2,260,000 *(91 per square mile)*
Language(s)	Latvian (*also known as* Lettish)
Area	24,942 *square miles*
Capital	**Riga** (REEG-uh)
Economy	$16,000 *annual per capita GDP*
Sovereignty	Democratic Republic *independence reestablished* 1991

LITHUANIA
(lith-oo-ANE-ee-uh)

Population	3,575,000 *(142 per square mile)*
Language(s)	Lithuanian
Area	25,213 *square miles*
Capital	**Vilnius** (VILL-nee-uss)
Economy	$15,300 *annual per capita GDP*
Sovereignty	Democratic Republic *independence reestablished* 1991

POLAND

Population	38,518,000 *(319 per square mile)*
Language(s)	Polish
Area	120,728 *square miles*
Capital	**Warsaw**
Economy	$14,400 *annual per capita GDP*
Sovereignty	Republic, *independence established* 1918

CZECH REPUBLIC
(check)

Population	10,229,000 *(336 per square mile)*
Language(s)	Czech
Area	30,450 *square miles*
Capital	**Prague** (prahg)
Economy	$22,000 *annual per capita GDP*
Sovereignty	Democratic Republic *independence established* 1993 (*the former* Czechoslovakia *became independent in* 1918)

SLOVAKIA
(sloh-VAH-kee-uh)

Population	5,448,000 *(288 per square mile)*
Language(s)	Slovak
Area	18,924 *square miles*
Capital	**Bratislava** (braht-ih-SLAHV-uh)
Economy	$18,200 *annual per capita GDP*
Sovereignty	Democratic Republic *independence established* 1993 (*the former* Czechoslovakia *became independent in* 1918)

HUNGARY

Population	9,956,000 *(277 per square mile)*
Language(s)	Hungarian
Area	35,919 *square miles*
Capital	**Budapest** (boo-duh-PESHT)
Economy	$17,500 *annual per capita GDP*
Sovereignty	Democratic Republic *independence established* 1001

ROMANIA

Population	22,276,000 *(243 per square mile)*
Language(s)	Romanian
Area	91,699 *square miles*
Capital	**Bucharest** (boo-kuh-REST)
Economy	$9,100 *annual per capita GDP*
Sovereignty	Republic, *independence established* 1878

BULGARIA

Population	7,323,000 *(171 per square mile)*
Language(s)	Bulgarian
Area	42,855 *square miles*
Capital	**Sofia**
Economy	$10,700 *annual per capita GDP*
Sovereignty	Democratic Republic *independence established* 1908

SLOVENIA
(sloh-VEEN-ee-uh)

Population	2,009,000 *(257 per square mile)*
Language(s)	Slovenian
Area	7,821 *square miles*
Capital	**Ljubljana** (Lee-oo-blee-AH-nah)
Economy	$23,400 *annual per capita GDP*
Sovereignty	Democratic Republic *independence established* 1991

CROATIA
(kroh-AY-shuh)

Population	4,493,000 *(206 per square mile)*
Language(s)	Croatian
Area	21,829 *square miles*
Capital	**Zagreb** (ZAH-grebb)
Economy	$13,400 *annual per capita GDP*
Sovereignty	Democratic Republic *independence established* 1991

BOSNIA and HERZEGOVINA
(BOZ-nee-uh *and* herts-uh-goh-VEEN-uh)

Population	4,552,000 *(230 per square mile)*
Language(s)	Croatian, Serbian, and Bosnian
Area	19,767 *square miles*
Capital	**Sarajevo** (sahr-uh-YAY-voh)
Economy	$5,600 *annual per capita GDP*
Sovereignty	Emerging Democratic Republic *independence established* 1992

SERBIA

Population	10,150,000 *(298 per square mile)*
Language(s)	Serbian
Area	34,116 *square miles*
Capital	**Belgrade**
Economy	$4,400 *annual per capita GDP*
Sovereignty	Republic, *independent since* 2006

MONTENEGRO
(mahn-tay-NAY-groh)

Population	685,000 *(128 per square mile)*
Language(s)	Montenegrin (a dialect of Serbian)
Area	5,333 *square miles*
Capital	**Podgorica** (POD-gor-eet-suh) (Podgorica *is the administrative capital;* Cetinje *is the other capital*)
Economy	$3,800 *annual per capita GDP*
Sovereignty	Republic, *independent since* 2006

MACEDONIA

Population	2,056,000 *(207 per square mile)*
Language(s)	Macedonian and Albanian
Area	9,928 *square miles*
Capital	**Skopje** (SKOPP-yuh)
Economy	$8,300 *annual per capita GDP*
Sovereignty	Democratic Republic *independent since* 1991

ALBANIA

Population	3,601,000 *(324 per square mile)*
Language(s)	Albanian
Area	11,100 *square miles*
Capital	**Tirana** (tih-RAHN-uh) *(also spelled* Tiranë)
Economy	$5,700 *annual per capita GDP*
Sovereignty	Emerging Democratic Republic *independence established* 1912

Review Exercises
for
Europe

Population Language Area Capital Economy Sovereignty

Country & Capital Identification

1. List the *capital city* of each of the 41 independent countries of Europe, and then indicate the location of each country on the map on the opposite page.

Country	Capital	Country	Capital
Albania		Lithuania	
Andorra		Luxembourg	
Austria		Macedonia	
Belgium		Malta	
Bosnia & Herzegovina		Monaco	
Bulgaria		Montenegro	
Croatia		Netherlands	
Cyprus		Norway	
Czech Republic		Poland	
Denmark		Portugal	
Estonia		Romania	
Finland		San Marino	
France		Serbia	
Germany		Slovakia	
Greece		Slovenia	
Hungary		Spain	
Iceland		Sweden	
Ireland		Switzerland	
Italy		United Kingdom	
Latvia		Vatican City	
Liechtenstein			

Population

2.   Which of the 41 independent countries of Europe has the *largest population*?
_____.

3.   Which of the 41 independent countries of Europe has the *smallest population*?
_____.

4.   Which of the 41 independent countries of Europe has the *highest population density*?
_____.

5.   Which of the 41 independent countries of Europe has the *lowest population density*?
_____.

Indicate the location of each of those countries on the map on the opposite page.

Area

6.   Which of the 41 independent countries of Europe has the *largest area*?
_____.

7.   Which of the 41 independent countries of Europe has the *smallest area*?
_____.

Indicate the location of each of those countries on the map on the opposite page.

Economy

8.   Which of the 41 independent countries of Europe has the *highest per capita GDP*?
_____.

9.   Which of the 41 sovereign states of Europe has the *lowest per capita GDP*?
_____.

Indicate the location of each of those countries on the map on the opposite page.

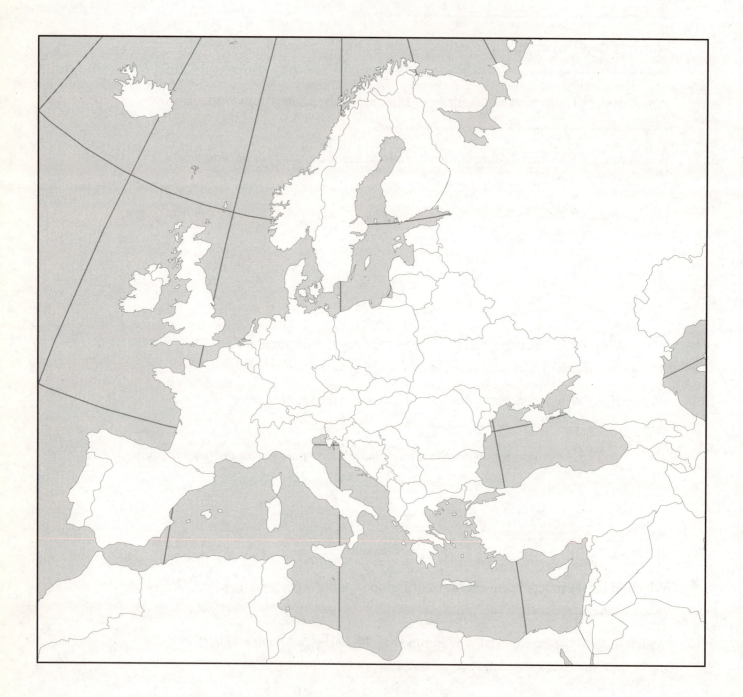

Languages

10. *French* and *German* are tied as the most widely used official or principal languages in Europe, with a total of six countries each. List the independent countries of Europe that have *French* or *German* as one of their official or principal languages:

	French			German	
1			1		
2			2		
3			3		
4			4		
5			5		
6			6		

Indicate the location of each of those countries on the map on the opposite page.

Sovereignty

11. Although the majority of Europe's countries are republics, a total of 10 monarchies can still be found in every part of the continent except Eastern Europe. List Europe's ten *monarchies*:

European Monarchies			
1		6	
2		7	
3		8	
4		9	
5		10	

Indicate the location of each of those countries on the map on the opposite page.

REGION 2
Russia & The Near Abroad

Russia and the Near Abroad

There are 12 independent countries in Russia and the Near Abroad:

Armenia	**Georgia**	**Moldova**	**Turkmenistan**
Azerbaijan	**Kazakhstan**	**Russia**	**Ukraine**
Belarus	**Kyrgyzstan**	**Tajikistan**	**Uzbekistan**

The 12 independent countries of Russia and the Near Abroad can be conveniently divided into three groups: the **Slavic States & Moldova**, the **Caucasus Region**, and **Central Asia**.

Slavic States & Moldova	Russia, Belarus, Ukraine, Moldova
Caucasus Region	Georgia, Armenia, Azerbaijan
Central Asia	Kazakhstan, Uzbekistan, Turkmenistan, Kyrgyzstan, Tajikistan

Russia and the Near Abroad consists of twelve of the fifteen former Soviet Socialist Republics that constituted the Union of Soviet Socialist Republics for the majority of the 20th century. The U.S.S.R. was officially disbanded on December 25, 1991, and replaced by fifteen independent countries with fifteen different principal languages. By a considerable margin, Russia was the largest and most important of the internal republics that made up the U.S.S.R. Not surprisingly, Russia inherited the permanent seat on the United Nations Security Council formerly held by the Soviet Union.

The U.S.S.R. has been succeeded by the Commonwealth of Independent States, which is comprised of 12 of the 15 members of the former Soviet Union (the Baltic States of Estonia, Latvia, and Lithuania, which were covered in *Region 1* on Europe, declined to join the CIS). The goals and purposes of the Commonwealth of Independent States include joint security and economic unity (see *Appendix B*).

Russia and the Near Abroad

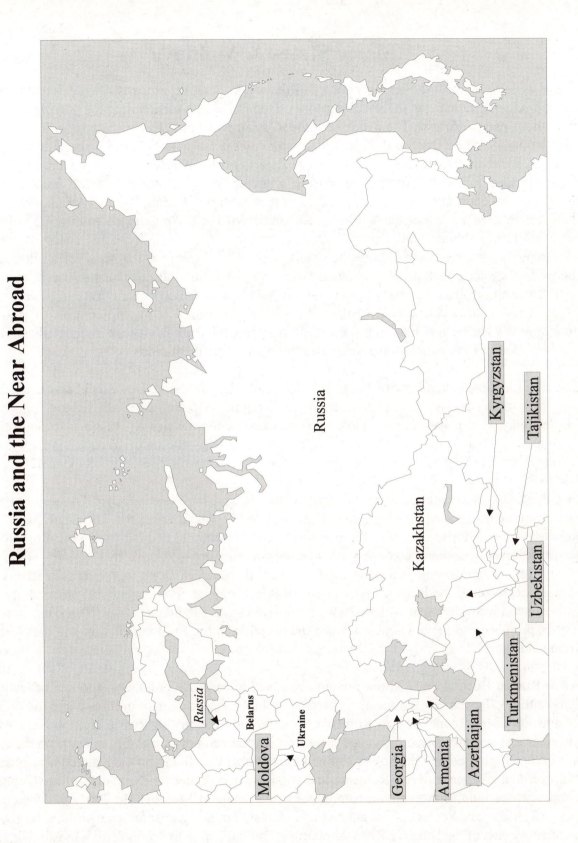

Russia

Kazakhstan

Kyrgyzstan

Tajikistan

Uzbekistan

Turkmenistan

Russia

Belarus

Ukraine

Moldova

Georgia

Armenia

Azerbaijan

Slavic States & Moldova

Russia is far and away the largest country in the world, with an immense territory that dominates the vast Eurasian landmass (notice that the small territory on the Baltic Sea surrounding the port city of Kaliningrad is a noncontiguous part of Russia). The majority of the Russian population lives in European Russia, to the west of the Ural Mountains, where most of the country's major cities can also be found. The eastern region of Russia between the Urals and the Pacific Ocean, which lies in Asia, is generally referred to as Siberia. Russia's national history can be traced to the 15th century, when the grand dukes of Moscow laid the foundation for what eventually became an enormous Tsarist empire. That empire collapsed during World War I and was succeeded by a socialist republic as a result of the Bolshevik Revolution of 1917. Although more than 4/5th of the present population is ethnically Russian, the country includes close to 100 other nationalities, many of whom live in autonomous republics scattered throughout the country (such as Chechnya, the scene of a bitter and bloody conflict between Russian troops and Chechnyans from 1994 to 1996, and the focus of a continuing guerrilla insurgency). Since the disbandment of the U.S.S.R., Russia has struggled in its efforts to build a democratic political system and market economy to replace the strict political and economic controls of the Soviet era. While some progress has been made on the economic front, recent years have seen a recentralization of power under President Vladimir Putin and an erosion in nascent democratic institutions.

The small landlocked country of Belarus is dominated by low-lying swampland and extensive forests; peat mining and timber are important elements of its economy. Belarus does not have a long tradition of national sovereignty, and it has retained closer political and economic ties to Russia than any other former Soviet republic. In 1999, Belarus and Russia signed a treaty on two-state union aimed at greater political and economic integration of the two countries, but the treaty has not yet been seriously implemented.

The third largest country in terms of area in Russia and the Near Abroad, Ukraine has the largest territory of any country located *entirely* inside the traditional boundaries of the European continent (which would exclude Russia and Turkey, both of which straddle the traditional boundary between Europe and Asia). Within those European confines, Ukraine's population size is surpassed only by Germany, France, United Kingdom, and Italy. Ukraine enjoys a wide range of natural resources along with a comparatively moderate climate. The breadbasket of the former Soviet Union, it is a leading area of agricultural productivity within the region. In a dramatic development in late 2004, the pro-Western candidate Viktor Yushchenko was elected president after promising to move Ukraine toward membership in the European Union and the North Atlantic Treaty Organization.

While Russia, Belarus, and Ukraine are closely linked historically, culturally, and linguistically (all three are Slavic countries), the Eastern European country of Moldova is an anomaly in Russia and the Near Abroad. Ethnically and linguistically identical to Romania, it is the only country in Russia and the Near Abroad that speaks a Romance (i.e., Latin-based) language. Moldova was carved out of territory taken from the eastern part of Romania and incorporated into the Soviet Union during World War II. Given their shared history and cultural identity, an eventual merger between Romania and Moldova would not be surprising, although reunification does not appear to be imminent given the current political climate in Moldova. In 2001, Moldova became the first former Soviet country to elect a Communist as its president. Russian forces remain in a narrow sliver of Moldovan territory east of the Dniester River, supporting the Slavic majority population, mostly Ukrainians and Russians, who have proclaimed a "Transnistria" republic.

Slavic States & Moldova

RUSSIA

Population	141,378,000 *(21 per square mile)*
Language(s)	Russian
Area	6,592,849 *square miles*
Capital	**Moscow** (MAHS-koh *or* MAHS-kow)
Economy	$12,200 *annual per capita GDP*
Sovereignty	Federal Republic, *independent republic since* 1917, *but origins of the Russian state can be traced to the* 12th *century Principality of Muscovy*

BELARUS
(BELL-uh-roos)

Population	9,725,000 *(121 per square mile)*
Language(s)	Belarusian
Area	80,155 *square miles*
Capital	**Minsk** (mensk)
Economy	$8,100 *annual per capita GDP*
Sovereignty	Republic *independent since* 1991

UKRAINE
(yoo-KRANE)

Population	46,300,000 *(199 per square mile)*
Language(s)	Ukrainian
Area	233,090 *square miles*
Capital	**Kiev** (KEE-yeff) *(also spelled Kyiv)*
Economy	$7,800 *annual per capita GDP*
Sovereignty	Republic *independence reestablished in* 1991

MOLDOVA
(mohl-DOH-vuh)

Population	4,320,000 *(331 per square mile)*
Language(s)	Moldovan (virtually the same as Romanian)
Area	13,070 *square miles*
Capital	**Chisinau** (kee-shee-NAU) [*formerly known as* Kishinev]
Economy	$2,000 *annual per capita GDP*
Sovereignty	Republic *independent since* 1991

Caucasus Region

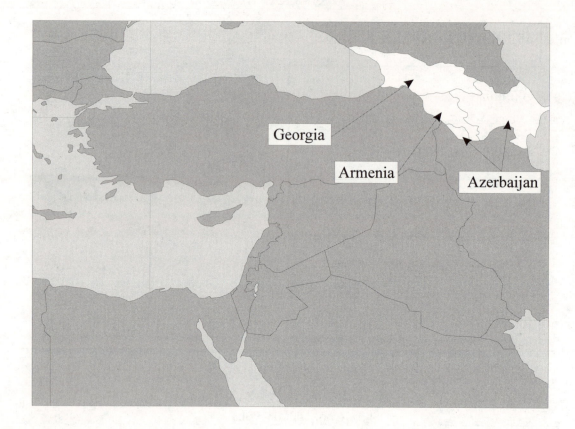

     The three former Soviet republics of Georgia, Armenia, and Azerbaijan are all dominated by the Caucasus Mountains, which connect the Caspian Sea and the Black Sea and form the traditional border between Asia and Europe. Georgia was absorbed into the Russian Empire in the 19th century, and then was briefly independent from 1918 to 1921 before being forcibly incorporated into the Soviet Union. Armenia, which was also part of the Russian Empire in the 19th century, was brought into the U.S.S.R. in 1920, as was Azerbaijan, which enjoyed a brief period of independence from 1918 to 1920.

     Today all three countries have predominantly agricultural economies. Georgia enjoys a subtropical region adjacent to the Black Sea that allows the production of citrus. Although primarily agrarian as well, Azerbaijan also produces petroleum from its Caspian Sea oil fields; corruption, however, is ubiquitous, and the country has not yet realized the widespread rise in its standard of living that its oil reserves would seem to promise. Armenia and Azerbaijan have been engaged in a long-standing and bitter territorial dispute over an Armenian enclave in Azerbaijan; despite a cease-fire in 1994, the dispute is still not completely resolved, and Azerbaijan is dealing with hundreds of thousands of refugees and internally displaced persons as a result of the conflict. Along with Georgia, Armenia's population is predominantly Orthodox Christian, whereas most of the people of Azerbaijan are Shi'ite Muslim. Notice that Azerbaijan has a noncontiguous piece of territory separated from the bulk of the country by an arm of Armenia.

GEORGIA

Population	4,646,000 *(173 per square mile)*
Language(s)	Georgian
Area	26,911 *square miles*
Capital	**T'bilisi** (tuh-BLEE-see)
Economy	$3,900 *annual per capita GDP*
Sovereignty	Republic *independence reestablished in* 1991

ARMENIA

Population	2,972,000 *(258 per square mile)*
Language(s)	Armenian
Area	11,506 *square miles*
Capital	**Yerevan** (yair-uh-VAHN)
Economy	$5,700 *annual per capita GDP*
Sovereignty	Republic *independence reestablished* 1991

AZERBAIJAN
(ah-zer-BIGH-jahn)

Population	8,120,000 *(243 per square mile)*
Language(s)	Azerbaijani
Area	33,437 *square miles*
Capital	**Baku** (bah-KOO) *(also spelled* Baki *or* Baky*)*
Economy	$7,500 *annual per capita GDP*
Sovereignty	Republic *independence reestablished in* 1991

Central Asia

Central Asia

The five independent countries of Central Asia have several features in common. The names of all five end in "stan," which is derived from the Persian word for "land of" or "home of." All five countries are landlocked (Kazakhstan and Turkmenistan border the Caspian Sea, but the Caspian Sea is actually a lake, with no outlet to the world ocean), and all five feature rugged terrains (Kazakhstan is a vast plateau, much of it desert, and large portions of the other countries are mountainous). Kazakhstan, Uzbekistan, Turkmenistan, Kyrgyzstan, and Tajikistan are all newly independent countries; unlike Afghanistan to the south, they were unable to resist Russian imperialism during the 20th century. All five countries are relatively poor, although all have potential wealth in the form of substantial mineral deposits (Kazakhstan, Uzbekistan, Turkmenistan, and Kyrgyzstan have particularly large reserves of natural gas, petroleum, coal, uranium, and other mineral ores). Finally, all five are, in varying proportions, substantially or predominantly Muslim countries.

The Aral Sea, on the border between Kazakhstan and Uzbekistan, was once the world's fourth largest lake, with abundant fish resources and a busy shipping trade. During the Soviet era, however, much of the water flowing into the Aral Sea was diverted to irrigate cotton fields, and the surface area of the lake shrank by about half. The sea level fell by more than 40 feet, and the salt content of the water doubled. The environmental consequences have been disastrous: dried salt on the exposed sea bottom is being carried away by the wind, poisoning the surrounding agricultural areas and creating an expanding desert around the lake.

KAZAKHSTAN
(kuh-ZAHK-stahn)

Population	15,285,000 *(15 per square mile)*
Language(s)	Kazakh & Russian
Area	1,049,156 *square miles*
Capital	**Astana** [*formerly known as* Akmola]
Economy	$9,400 *annual per capita GDP*
Sovereignty	Republic *independent since* 1991

UZBEKISTAN
(ooz*-BECK-ih-stahn [* "oo" as in "wood"])

Population	27,780,000 *(161 per square mile)*
Language(s)	Uzbek
Area	172,742 *square miles*
Capital	**Tashkent** (tahsh-KENT) *(sometimes spelled Toshkent)*
Economy	$2,000 *annual per capita GDP*
Sovereignty	Republic *independent since* 1991

TURKMENISTAN
(turk-menn-ih-stahn)

Population	5,097,000 *(27 per square mile)*
Language(s)	Turkmen
Area	188,457 *square miles*
Capital	**Ashgabat** (ASH-gah-baht) *(also spelled* Ashkhabad)
Economy	$8,500 *annual per capita GDP*
Sovereignty	Republic *independent since* 1991

KYRGYZSTAN
(keer-geez-stahn)

Population	5,284,000 *(68 per square mile)*
Language(s)	Kyrgyz & Russian
Area	77,182 *square miles*
Capital	**Bishkek**
Economy	$2,100 *annual per capita GDP*
Sovereignty	Republic *independent since* 1991

TAJIKISTAN
(tah-JEEK-ih-stahn)

Population	7,077,000 *(128 per square mile)*
Language(s)	Tajik & Russian
Area	55,251 *square miles*
Capital	**Dushanbe** (doo-SHAN-bay)
Economy	$1,300 *annual per capita GDP*
Sovereignty	Republic *independent since* 1991

Review Exercises
for
Russia & the Near Abroad

Population Language Area Capital Economy Sovereignty

Country & Capital Identification

1. List the *capital city* of each of the 12 independent countries of Russia & the Near Abroad, and then indicate the location of each country on the map on the opposite page.

Country	Capital
Armenia	
Azerbaijan	
Belarus	
Georgia	
Kazakhstan	
Kyrgyzstan	
Moldova	
Russia	
Tajikistan	
Turkmenistan	
Ukraine	
Uzbekistan	

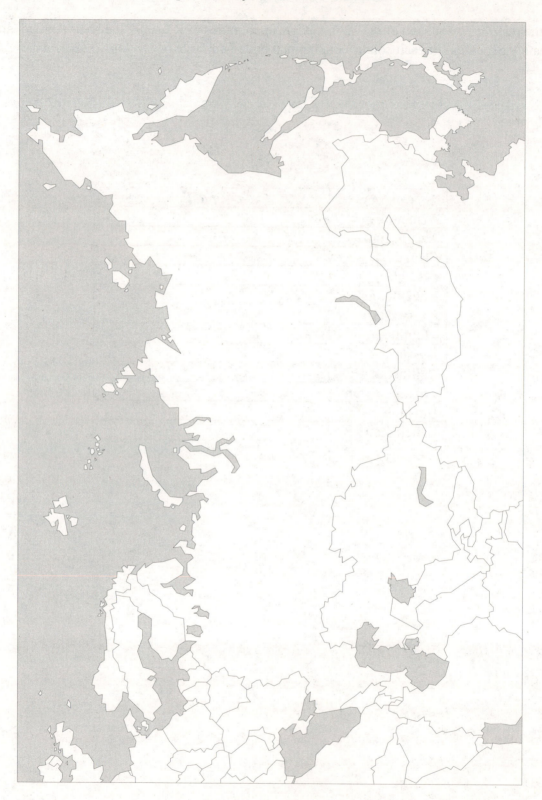

Population

2. With more than 140 million people, Russia has by far the largest population of any of the 12 countries in Russia and the Near Abroad. Which of the 12 independent countries in Russia & the Near Abroad has the *second largest population*? _____

3. Which of the 12 independent countries in Russia & the Near Abroad has the *smallest population*?

   _____

4. Which of the 12 independent countries in Russia & the Near Abroad has the *highest population density*?

   _____

5. Which of the 12 independent countries in Russia & the Near Abroad has the *lowest population density?*

   _____

   Indicate the location of each of those countries on the map on the opposite page.

Area

6. Russia has by far the largest area of any of the 12 countries in Russia & the Near Abroad (in fact, Russia is substantially larger than the other 11 combined). Which country has the *second largest area?*

   _____

7. Which of the 12 independent countries in Russia & the Near Abroad has the *smallest area?*

   _____

   Indicate the location of each of those countries on the map on the opposite page.

Economy

8. Which of the 12 independent countries in Russia & the Near Abroad has the *highest per capita GDP?*

   _____

9. Which of the 12 independent countries in Russia & the Near Abroad has the *lowest per capita GDP?*

   _____

   Indicate the location of each of those countries on the map on the opposite page.

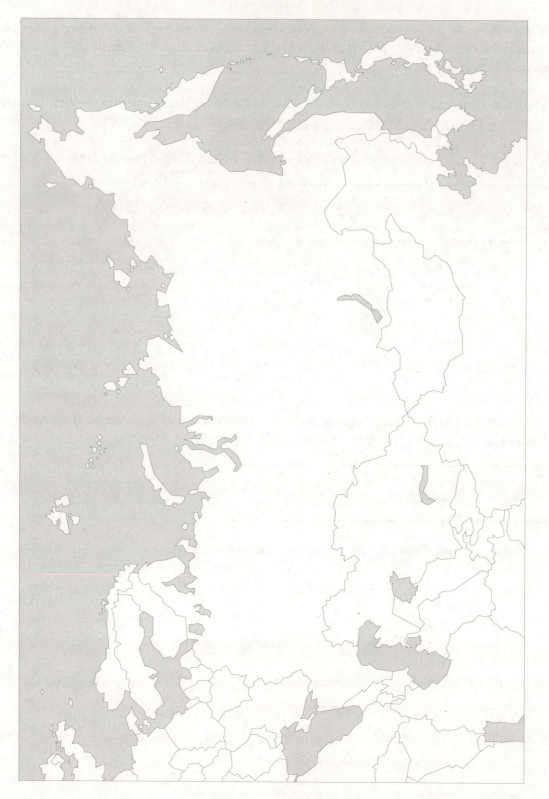

Languages

10.   Name the only independent country in Russia & the Near Abroad that has a *Romance (i.e., Latin-based) language* as its principal language: _____

        Indicate the location of that country on the map on the opposite page.

Notice that the language of each of the 12 countries in Russia and the Near Abroad is derived from the name of the country itself; thus if you can name and locate each of the 12 countries, you can name each of the 12 principal languages described in this chapter, and identify the location of the country where each is spoken.

Sovereignty

11.   Name the only federal republic among the 12 independent countries in Russia & the Near Abroad:

_____

        Indicate the location of that country on the map on the opposite page.

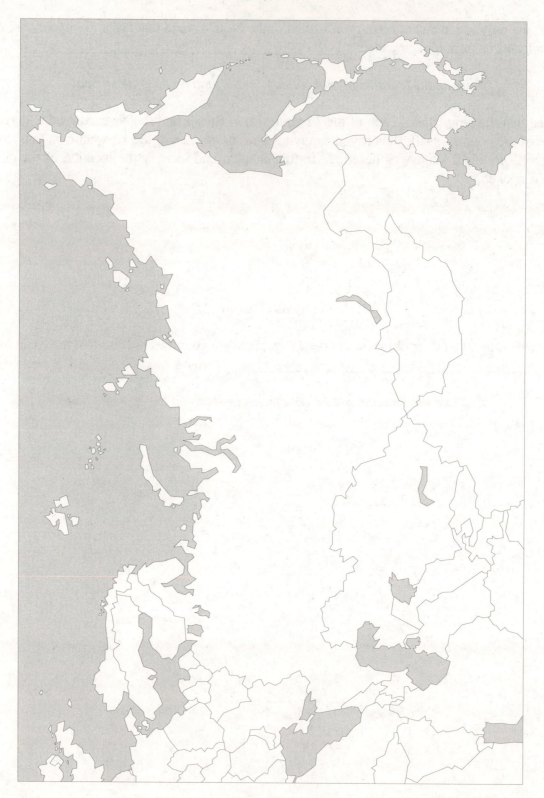

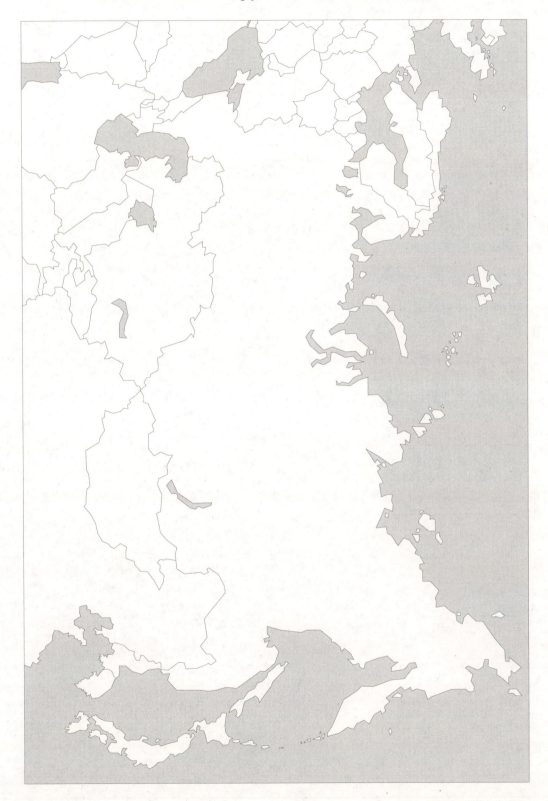

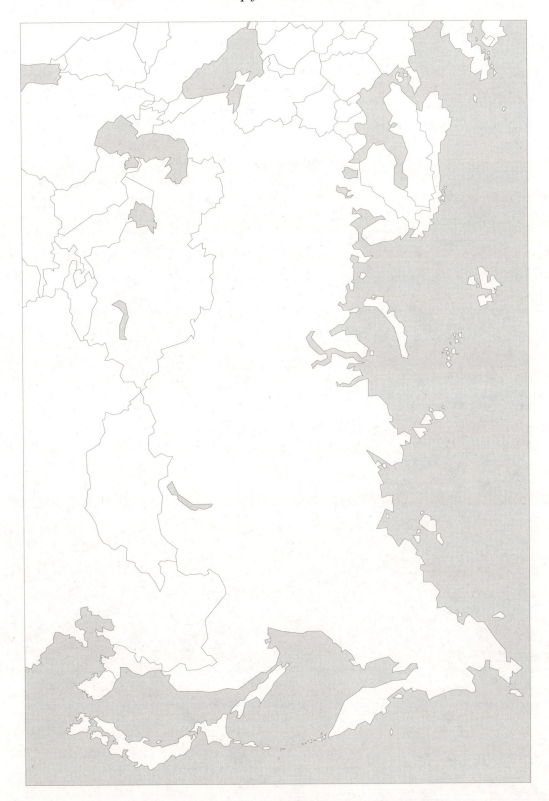

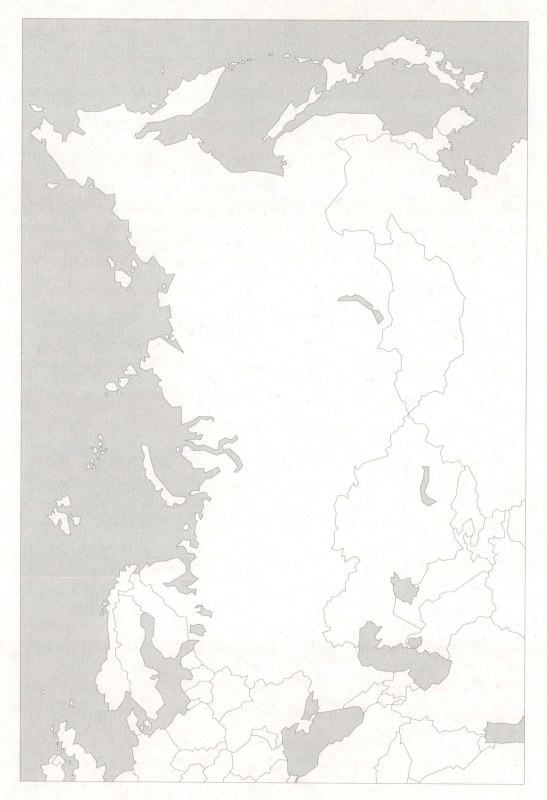

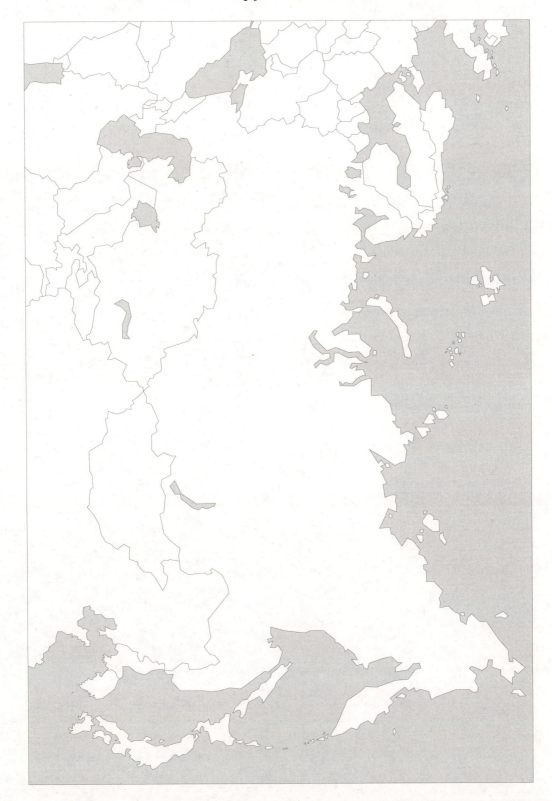

REGION 3
Middle East & North Africa

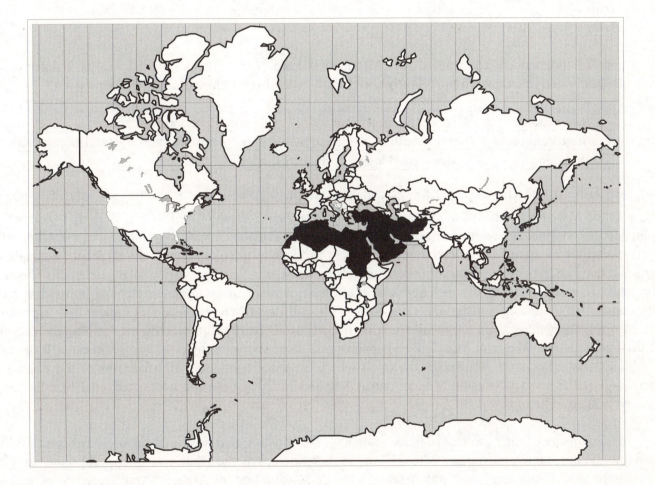

The Middle East and North Africa

There are 21 independent countries in the Middle East and North Africa:

Afghanistan	**Israel**	**Morocco**	**Syria**
Algeria	**Jordan**	**Oman**	**Tunisia**
Bahrain	**Kuwait**	**Qatar**	**Turkey**
Egypt	**Lebanon**	**Saudi Arabia**	**United Arab Emirates**
Iran	**Libya**	**Sudan**	**Yemen**
Iraq			

The 21 independent countries of the Middle East and North Africa covered in this chapter can be conveniently divided into three groups: **Southwest Asia**, the **Arabian Peninsula**, and **North Africa**.

Southwest Asia	Turkey, Iran, Afghanistan, Iraq, Syria, Lebanon, Jordan, Israel
Arabian Peninsula	Saudi Arabia, Yemen, Oman, United Arab Emirates, Qatar, Bahrain, Kuwait
North Africa	Morocco, Algeria, Tunisia, Libya, Egypt, Sudan

The **Middle East** is a Eurocentric term whose origins date from the days when Europeans thought of Turkey as the *Near East*, India as the *East*, and China as the *Far East*; the *Middle East*, from their perspective, designated the intermediate region surrounding the Persian Gulf. Today, as a geographic term, the *Middle East* can have various meanings. Most commonly, perhaps, it refers to the countries of the Arabian Peninsula and their immediate neighbors on the eastern shore of the Mediterranean Sea. The region identified in this chapter as the *Middle East & North Africa* includes a 6,000-mile-wide span of countries stretching from Morocco in northwestern Africa to Afghanistan in central Asia. Although the various peoples that inhabit this large region are far from homogenous, there are many common threads of language, culture, religion, history, and economy that tie them together. The term *Middle East* is appropriate in at least one important sense, because it describes the region's global location. Situated at the point where the three continents of Africa, Asia, and Europe are joined (and at the point where the Mediterranean Sea and Indian Ocean are connected), the Middle East is arguably in the middle of the world.

The majority of the countries in the Middle East are **Muslim**, and most of them are **Arabic** as well, but the two terms are *not* synonymous. "Muslim" refers to followers of the Islamic religion, and there are many people in the world, including most Indonesians, who are Muslim but not Arabic. "Arab" refers to people who share a particular ethnic and cultural identity, most readily distinguished by the fact that they all speak the Arabic language. As it happens, all Arab countries are predominantly Muslim, but not all Muslim countries are predominantly Arab. The Arab world extends across North Africa and into Southwest Asia, and includes Morocco, Algeria, Tunisia, Libya, Egypt, Syria, Lebanon, Jordan, and Iraq, as well as all the countries of the Arabian Peninsula. However, the Arab world does not include Turkey, Iran, Afghanistan, or Pakistan, all of which are Islamic but none of which speak Arabic.

North Africa

Morocco

Tunisia

Algeria

Libya

Egypt

Western Sahara
(Morocco)

Sudan

The Middle East

Turkey

Syria

Lebanon

Kuwait

Afghanistan

Iraq

Iran

Israel

Jordan

United Arab Emirates

Bahrain

Qatar

Oman

Saudi Arabia

Yemen

Southwest Asia

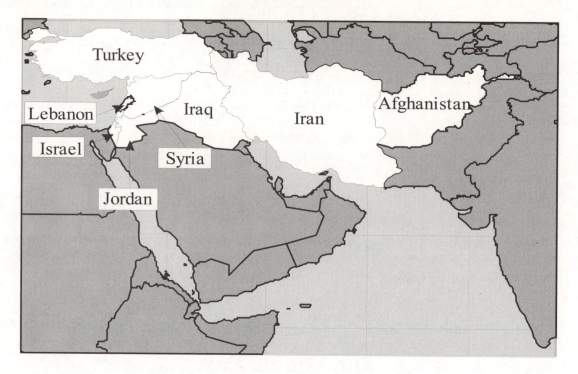

Southwest Asia was home to the world's first state society (the Sumerian civilization, in what is today the country of Iraq), and it is the birthplace of three major world religions (Judaism, Christianity, and Islam). Today, the only country in Southwest Asia that is not predominantly Muslim is Israel, which is predominantly Jewish but which does have a growing Muslim minority.

The great majority of the large country of Turkey lies in Asia, but a small portion of Turkish territory on the western side of the Sea of Marmara lies within the traditional boundaries of Europe. Although it is a Muslim country, Turkey is the most westernized of the Islamic states in the Middle East. Its location at the crossroads between Europe and Asia has contributed much to the development of Turkey's national identity. The modern country of Turkey is the successor to the imperialistic Ottoman Empire, which, for several centuries until World War I, dominated much of the Balkan Peninsula. The country of Bulgaria still has a significant Turkish minority, as does Germany.

The bitter ongoing conflict between the Israelis and Palestinians defies easy summary, because partisans on both sides are likely to object strenuously to every omission and to argue vehemently about every nuance of the account. What follows, therefore, should be regarded as only the sketchiest overview. The state of Israel was established in 1948 on the recommendation of a United Nations commission, although much of the territory of Israel was already occupied by Palestinian Arabs. The surrounding Arab states of Egypt, Syria, Lebanon, and Jordan responded by immediately invading the new country of Israel, but were quickly driven back (aided in no small part by the superior weaponry provided to Israel by the Western democracies, including the United States). In 1964, the Palestine Liberation Organization was founded with the declared intent of reclaiming the land taken from the Palestinian Arabs and destroying the state of Israel.

In 1967, in the second Arab-Israeli war (the Six-Day War), Israel seized control of the Sinai peninsula from Egypt, the Golan Heights from Syria, and the West Bank and the Old City of Jerusalem from Jordan.

(Contemporary Palestinian demands for a return to the "pre-1967" borders refer to these territories.) In 1973, in the third Arab-Israeli war (the Yom Kippur War), Israel defeated the allied forces of Egypt and Syria and solidified control of its captured territory. In 1979, Egyptian President Anwar Sadat signed a peace treaty with Israel; the terms of the treaty included a mutual recognition pact between Egypt and Israel and the return of the Sinai peninsula to Egypt. Sadat was subsequently assassinated by Arab extremists opposed to any peace with Israel.

In 1981, Israel formally annexed the Golan Heights. The Arabs responded with terrorist attacks along Israel's northern border, and Israel responded by invading Lebanon in 1982. The Israelis withdrew from most of Lebanon in 1985, but they maintained a "security zone" along the border, so Syria stationed troops in Lebanon. In 1987, the Palestinians launched the *intifada*, or popular uprising, in response to the systematic establishment of Jewish settlements on Palestinian land in the West Bank.

With the Oslo Accords of 1993, Israelis (under the leadership of Prime Minister Yitzhak Rabin) and Palestinians (represented by the Palestine Liberation Organization, under the leadership of Yasser Arafat) reached a major agreement in which the PLO recognized Israel's right to exist, while Israel for its part pledged to grant Palestinians limited self-rule in the West Bank and Gaza. Rabin was assassinated in 1995 by a Jewish extremist opposed to any peace with the Palestinians. Rabin's death inspired calls for a rededication to the peace process on the part of several Arab leaders. In 1998, Israel and the PLO signed the Wye River Memorandum, which worked out some of the details for implementing the 1993 accords.

Putting the peace into practice proved to be insurmountably difficult, however. In late 2000, violence flared in the West Bank and Gaza after Israeli politician Ariel Sharon (who had led the 1982 invasion of Lebanon) made a provocative visit to the Haram al-Sharif mosque compound in Jerusalem. (The ultimate disposition of Jerusalem, which Moslems, Jews, and Christians all regard as a holy city, was one of the unresolved issues in the Oslo Accords.) Sharon became prime minister in 2001, and following the terrorist attacks on the United States on September 11th, Israel escalated its own "war on terrorism." In an incendiary spiral of violence, Palestinians launched a new wave of suicide bomb attacks, and Israel responded with military occupation of the West Bank. Many lives were lost on both sides.

A watershed event in Middle Eastern affairs occurred in November 2004 when Palestinian President Yasser Arafat died. Seizing the opportunity for a new beginning, Israelis and Palestinians agreed to a cease-fire at a summit at Sharm al-Sheikh in February 2005, and Israel disengaged from the Gaza strip that summer. The next several months were marked by renewed violence, however, and Israel shelved plans to disengage from the West Bank after HAMAS gained control of Gaza (Israel considers HAMAS to be a terrorist organization). In November 2007, Palestinian President Mahmoud Abbas and Israeli Prime Minister Ehud Olmert met at the U.S. Naval Academy in Annapolis, Maryland, and announced a joint agreement to reach a peace pact by the end of 2008. The two parties pledged to establish "a democratic Palestinian state that will live side by side with Israel in peace and security," but the details of the agreement remain to be negotiated, and the ultimate fate of both Jerusalem and Gaza are formidable obstacles to be overcome in the struggle for a permanent peace agreement.

TURKEY

Population	71,159,000 *(235 per square mile)*
Language(s)	Turkish *plus* Kurdish & Armenian
Area	302,541 *square miles*
Capital	**Ankara** (ANG-kah-ruh)
Economy	$9,100 *annual per capita GDP*
Sovereignty	Democratic Republic *independent since* 1923

IRAN
(ih-RAN *or* ih-RAHN)

Population	65,398,000 *(103 per square mile)*
Language(s)	Farsi (*i.e.*, Persian)
Area	636,372 *square miles*
Capital	**Tehran** (teh-RAHN)
Economy	$8,700 *annual per capita GDP*
Sovereignty	Theocratic Republic *independent since* 1979

AFGHANISTAN

Population	31,890,000 *(127 per square mile)*
Language(s)	Pashtu, Dari, Uzbek, & Turkmen *plus many others*
Area	251,773 *square miles*
Capital	**Kabul** (KAH-bull)
Economy	$800 *annual per capita GDP*
Sovereignty	Republic *independent since* 1919

IRAQ
(ih-RAK *or* ih-RAHK)

Population	27,500,000 *(162 per square mile)*
Language(s)	Arabic & Kurdish
Area	169,235 *square miles*
Capital	**Baghdad**
Economy	$1,900 *annual per capita GDP*
Sovereignty	Transitional Democratic Republic *independent since* 1932

SYRIA
(SEARY-uh)

Population	19,315,000 *(270 per square mile)*
Language(s)	Arabic *plus* Kurdish, Armenian, Aramaic, & Circassian
Area	71,498 *square miles*
Capital	**Damascus** (duh-MASK-uss)
Economy	$4,100 *annual per capita GDP*
Sovereignty	Republic *independent since 1946*

LEBANON

Population	3,926,000 *(978 per square mile)*
Language(s)	Arabic *plus* French, English, & Armenian
Area	4,016 *square miles*
Capital	**Beirut** (bay-ROOT)
Economy	$5,900 *annual per capita GDP*
Sovereignty	Republic *independent since 1943*

JORDAN

Population	6,053,000 *(175 per square mile)*
Language(s)	Arabic
Area	34,495 *square miles*
Capital	**Amman** (uh-MAHN)
Economy	$5,100 *annual per capita GDP*
Sovereignty	Constitutional Monarchy *independent since* 1946

ISRAEL

Population	6,427,000 *(801 per square mile)*
Language(s)	Hebrew & Arabic
Area	8,019 *square miles*
Capital	**Jerusalem** *(most embassies are in Tel Aviv)*
Economy	$26,800 *annual per capita GDP*
Sovereignty	Democratic Republic *independent since* 1948

Arabian Peninsula

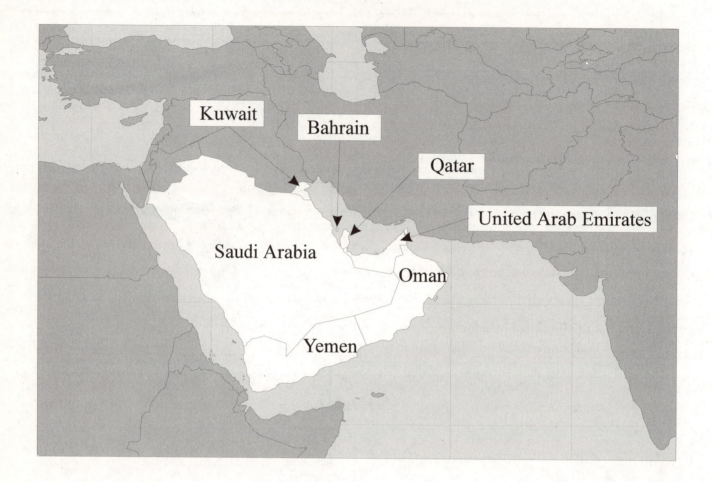

The large, sparsely-populated Arabian Peninsula, which is almost entirely covered by a harsh desert, has few natural resources, with one monumental exception: it has the largest known concentrations of petroleum reserves in the world. Saudi Arabia occupies the bulk of the peninsula, and accounts for the majority of its population as well. Mecca, the holiest city in the Islamic religion and the goal of pilgrimages for Muslims worldwide, is located in Saudi Arabia.

The second most populous country on the Arabian Peninsula, Yemen, is also the poorest, but recently discovered oil reserves may soon raise its petroleum exports to the levels of its neighbors. The tiny country of Bahrain consists of 35 islands lying close offshore Saudi Arabia. Like Qatar, Oman, and the United Arab Emirates, Bahrain is a major petroleum exporter, but the archipelago country also generates income as an offshore banking center. Notice that Oman has a noncontiguous piece of territory on the peninsula that overlooks the Strait of Hormuz; much of the oil consumed in western Europe, the United States, and Japan is carried in supertankers past this strategic chokepoint.

SAUDI ARABIA

Population	27,601,000 *(33 per square mile)*
Language(s)	Arabic
Area	830,000 *square miles*
Capital	**Riyadh** (ree-YADH)
Economy	$13,800 *annual per capita GDP*
Sovereignty	Monarchy *independent since* 1932

YEMEN
(YEM-MON)

Population	22,231,000 *(109 per square mile)*
Language(s)	Arabic
Area	203,850 *square miles*
Capital	**Sanaa** (sah-NAH)
Economy	$1,000 *annual per capita GDP*
Sovereignty	Republic *independent since* 1990

OMAN
(oh-MAHN)

Population	3,205,000 *(39 per square mile)*
Language(s)	Arabic
Area	82,030 *square miles*
Capital	**Muscat** (MUH-skaht)
Economy	$14,400 *annual per capita GDP*
Sovereignty	Monarchy *independence established* 1650

UNITED ARAB EMIRATES

Population	4,444,000 *(138 per square mile)*
Language(s)	Arabic
Area	32,278 *square miles*
Capital	**Abu Dhabi** (AH-boo dah-bee) [*also known as* Abu Zaby]
Economy	$49,700 *annual per capita GDP*
Sovereignty	Federation of Emirates *independent since* 1971

QATAR
(KOT-TER)

Population	907,000 *(206 per square mile)*
Language(s)	Arabic
Area	4,412 *square miles*
Capital	**Doha** (DOH-hah)
Economy	$29,800 *annual per capita GDP*
Sovereignty	Monarchy *independent since* 1971

BAHRAIN

Population	709,000 *(2,655 per square mile)*
Language(s)	Arabic
Area	267 *square miles*
Capital	**Manama** (MAH-nah-mah)
Economy	$25,600 *annual per capita GDP*
Sovereignty	Constitutional Monarchy *independent since* 1971

KUWAIT

Population	2,506,000 *(364 per square mile)*
Language(s)	Arabic
Area	6,880 *square miles*
Capital	**Kuwait City**
Economy	$23,100 *annual per capita GDP*
Sovereignty	Constitutional Monarchy *independent since* 1961

North Africa

The five North African countries that border the Mediterranean coast are all Arabic-speaking Muslim countries, as is Sudan to the south. As a cultural region, the northern tier of African countries belongs to the Arab World. The Arab and Islamic influences have made significant inroads into the second tier of North African countries as well. The southern portion of the Sahara Desert is often referred to as the African Transition Zone because its cultures reflect a blend of Arab and sub-Saharan characteristics. The Arabic language is spoken in several of those countries, and Muslims make up significant portions of the populations.

North Africa includes the continent's first and second largest countries in terms of geographic area, Sudan and Algeria. The second most populous country in Africa, Egypt, is also found in North Africa. There is a major unresolved territorial dispute in the region at the moment involving the area of Western Sahara, which was known as Spanish Sahara until Spain relinquished control in 1975. The territory has been under effective Moroccan control since 1979, when Mauritania abandoned a competing claim. Since 1991 a United Nations Mission in Western Sahara has been working to reach a resolution of the dispute. The UN Mission is hoping to hold a referendum that would allow the people of Western Sahara to express their preference regarding sovereignty, but to date no agreement has been reached on the timing or the structure of such a referendum.

North Africa

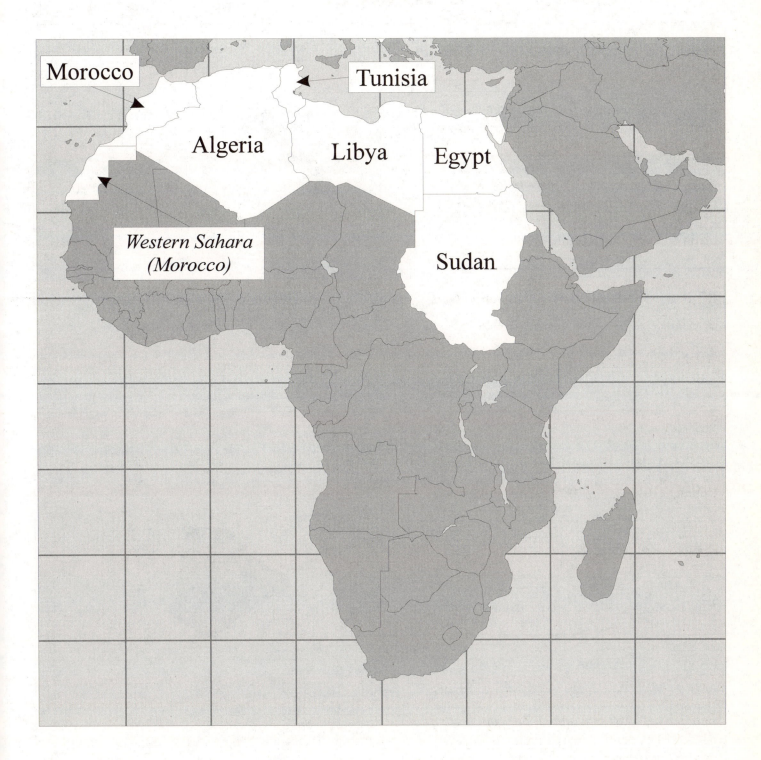

MOROCCO

Population	33,757,000 *(123 per square mile)*
Language(s)	Arabic (*although* French *is often the language of business & government*)
Area	275,117 *square miles (including Western Sahara)*
Capital	**Rabat** (rah-BAHT)
Economy	$4,600 *annual per capita GDP*
Sovereignty	Constitutional Monarchy *independent since* 1956

ALGERIA

Population	33,333,000 *(36 per square mile)*
Language(s)	Arabic
Area	919,595 *square miles*
Capital	**Algiers**
Economy	$7,600 *annual per capita GDP*
Sovereignty	Republic *independent since* 1962

TUNISIA
(too-NEE-zhuh)

Population	10,276,000 *(163 per square mile)*
Language(s)	Arabic & French
Area	63,170 *square miles*
Capital	**Tunis** (TOO-niss)
Economy	$8,900 *annual per capita GDP*
Sovereignty	Republic *independent since 1956*

LIBYA

Population	6,037,000 *(9 per square mile)*
Language(s)	Arabic
Area	679,362 *square miles*
Capital	**Tripoli** (TRIPPA-lee)
Economy	$12,300 *annual per capita GDP*
Sovereignty	Republic *independent since 1951*

EGYPT

Population	80,335,000 *(208 per square mile)*
Language(s)	Arabic
Area	386,662 *square miles*
Capital	**Cairo** (KIGH-roh) [Al-Qahira *in Arabic*]
Economy	$4,200 *annual per capita GDP*
Sovereignty	Republic *independent since 1922*

SUDAN

Population	39,379,000 *(41 per square mile)*
Language(s)	Arabic & English *plus several indigenous languages*
Area	967,500 *square miles*
Capital	**Khartoum** (kahr-TOOM)
Economy	$2,400 *annual per capita GDP*
Sovereignty	Republic *independent since 1956*

Review Exercises
for
The Middle East & North Africa

Population Language Area Capital Economy Sovereignty

Country & Capital Identification

1. List the *capital city* of each of the 21 independent countries of the Middle East & North Africa, and then indicate the location of each country on the maps on the opposite page.

Country	Capital
Afghanistan	
Algeria	
Bahrain	
Egypt	
Iran	
Iraq	
Israel	
Jordan	
Kuwait	
Lebanon	
Libya	
Morocco	
Oman	
Qatar	
Saudi Arabia	
Sudan	
Syria	
Tunisia	
Turkey	
U. A. Emirates	
Yemen	

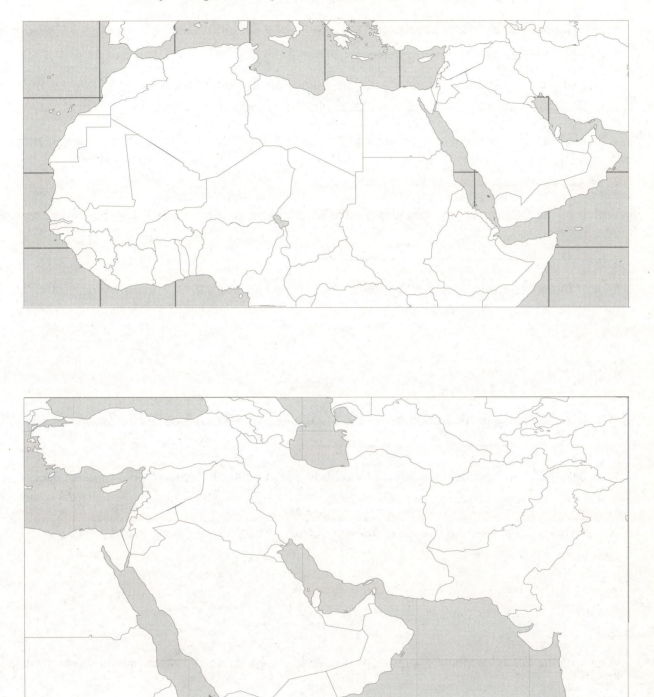

Population

2. Which of the 21 independent countries of the Middle East & North Africa has the *largest population*?

    _____

3. Which of the 21 independent countries of the Middle East & North Africa has the *smallest population*?

    _____

4. Which of the 21 independent countries of the Middle East & North Africa has the *highest population density*?

    _____

5. Which of the 21 independent countries of the Middle East & North Africa has the *lowest population density*?

    _____

    Indicate the location of each of those countries on the maps on the opposite page.

Area

6. Which of the 21 independent countries of the Middle East & North Africa has the *largest area*?

    _____

7. Which of the 21 independent countries of the Middle East & North Africa has the *smallest area*?

    _____

    Indicate the location of each of those countries on the maps on the opposite page.

Economy

8. Which of the 21 independent countries of the Middle East & North Africa has the *highest per capita GDP?*

    _____

9. Which of the 21 independent countries of the Middle East & North Africa has the *lowest per capita GDP?*

    _____

    Indicate the location of each of those countries on the maps on the opposite page.

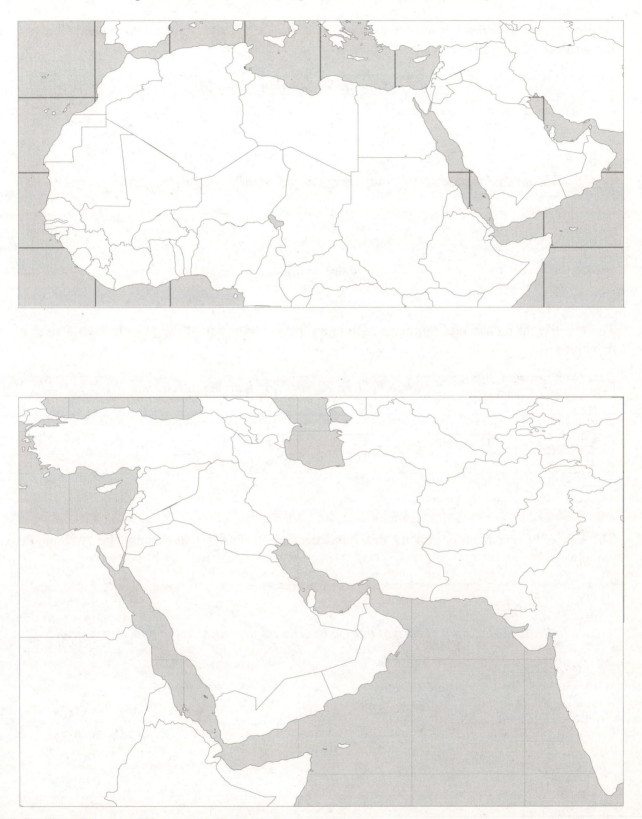

Languages

10. Of the 21 independent countries in the Middle East & North Africa, three do *not* have Arabic as one of their official or principal languages. Name those three countries:

Non-Arabic Speaking Countries			
1		3	
2			

Indicate the location of each of those countries on the maps on the opposite page.

Sovereignty

11. There are eight monarchies among the 21 independent countries of the Middle East & North Africa, namely:

Middle Eastern & North African Monarchies			
1		5	
2		6	
3		7	
4		8	

12. There is only one republic among the 7 independent countries that occupy the Arabian Peninsula, namely:

_____

Indicate the location of each of those countries on the maps on the opposite page.

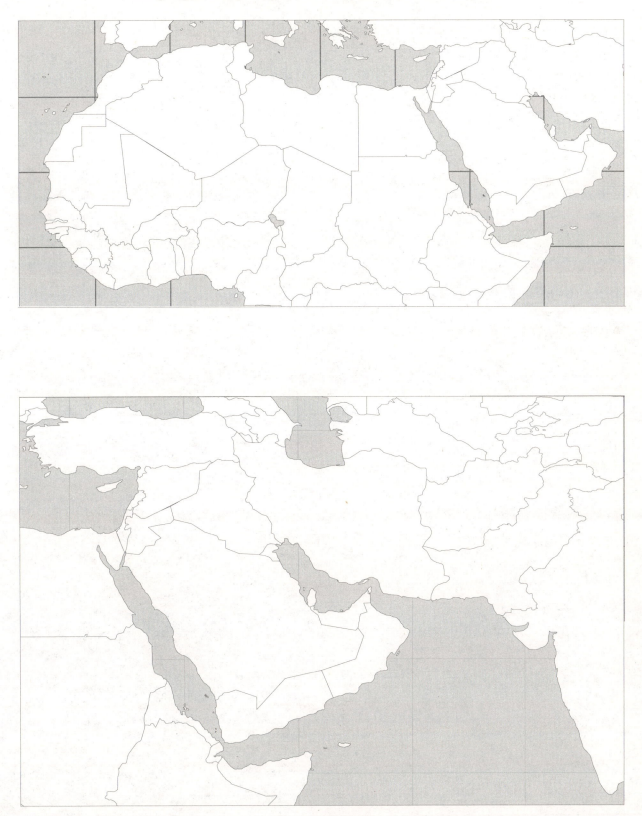

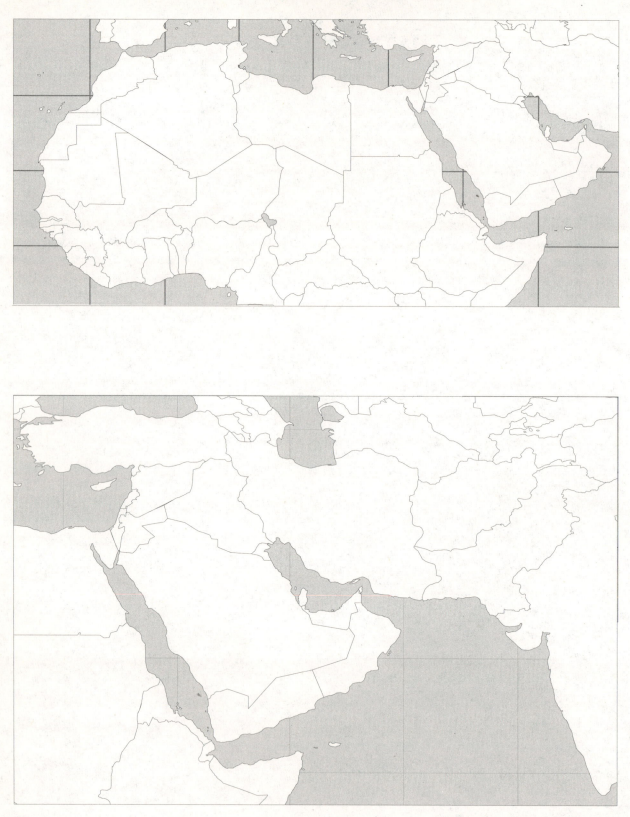

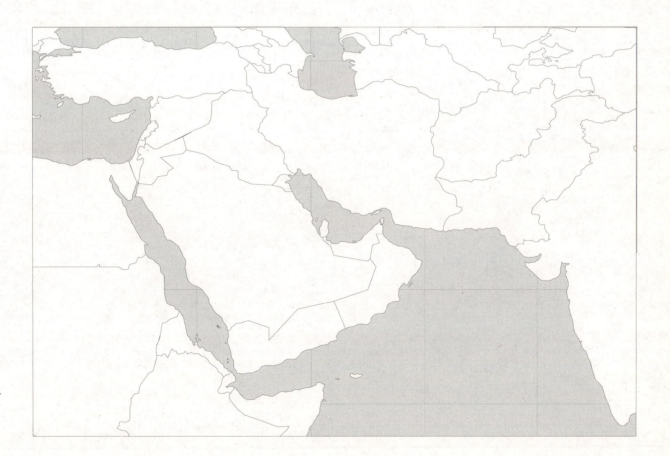

Places of the World

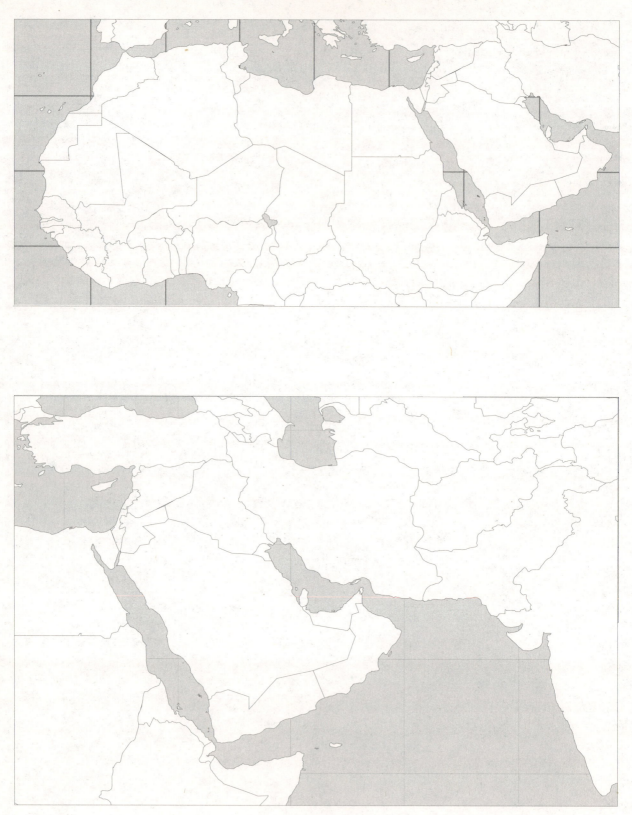

Places of the World

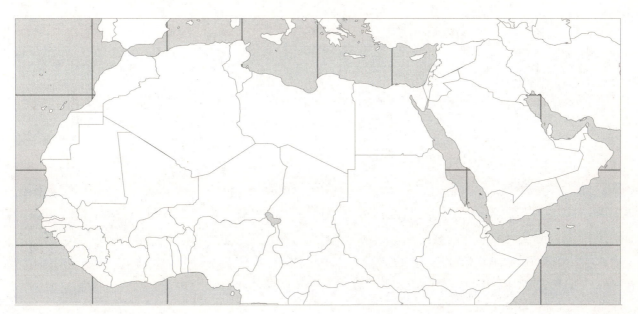

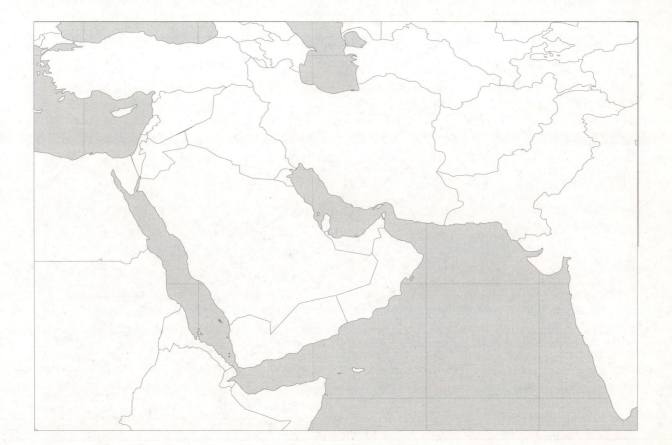

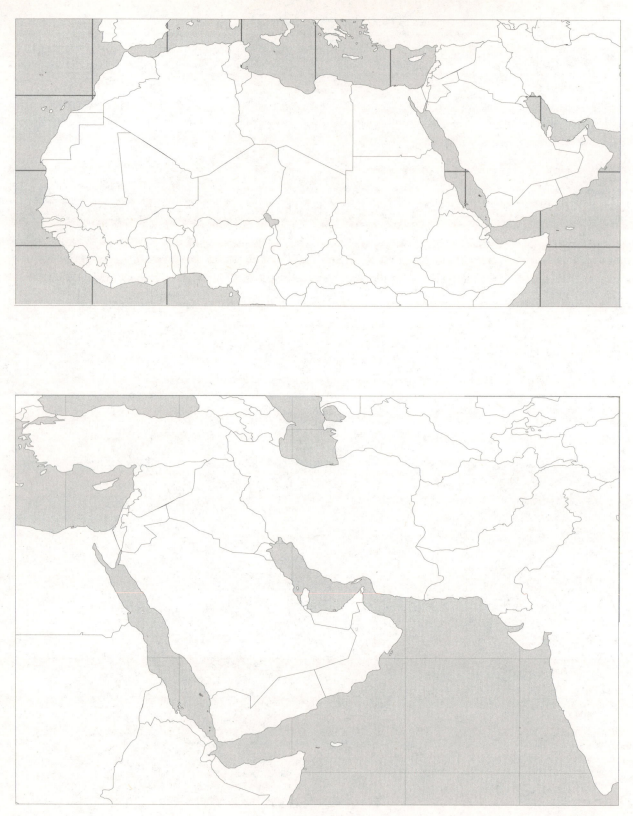

REGION 4
Monsoon Asia

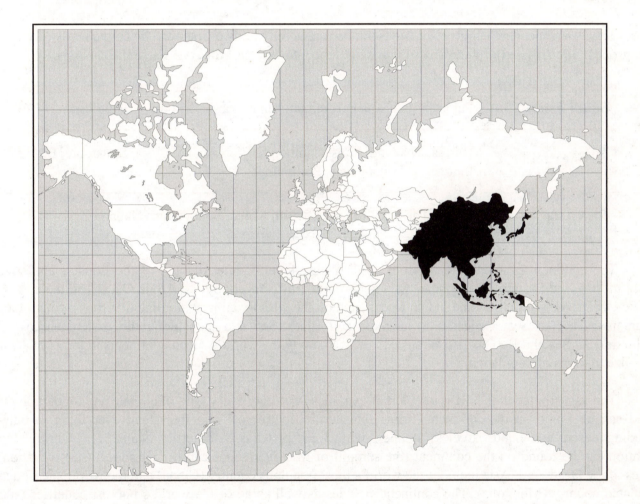

Monsoon Asia

There are 24 independent countries in Monsoon Asia:

Bangladesh	India	Mongolia	Sri Lanka
Bhutan	Indonesia	Nepal	Taiwan
Brunei	Japan	North Korea	Thailand
Burma	Laos	Pakistan	Timor-Leste
Cambodia	Malaysia	Philippines	Vietnam
China	Maldives	South Korea	Singapore

The 24 independent countries of Monsoon Asia covered in this chapter can be conveniently divided into three groups: **South Asia** (also known as the Indian Subcontinent), **Southeast Asia**, and **East Asia**.

South Asia	Pakistan, India, Nepal, Bhutan, Bangladesh, Sri Lanka, Maldives
Southeast Asia	Burma, Thailand, Laos, Cambodia, Vietnam, Malaysia, Singapore, Brunei, Indonesia, Timor-Leste, Philippines
East Asia	China, Taiwan, Mongolia, North Korea, South Korea, Japan

The continent of **Asia** extends from north of the Arctic Circle to south of the Equator and stretches across most of the Eastern Hemisphere. Asia is bordered on the north by the Arctic Ocean, on the east by the Pacific Ocean, on the south by the Indian Ocean, and on the west by an irregular line that runs through the Ural Mountains, Caspian Sea, Caucasus Mountains, and Black Sea and then along the eastern shores of the Mediterranean Sea and Red Sea. Asia ranks 1st in size among the world's seven continents in terms of area, and 1st in size in terms of population.

The vast continent of Asia is a region of extremes. Virtually every kind of climate and topography in the world can be found on the continent, from arctic tundra to tropical rain forest and everything in between, including desert wastelands, treeless steppes, windswept plateaus, and fertile plains. The world's highest elevation can be found on the continent (the summit of Mt. Everest) as can the world's lowest elevation (the shores of the Dead Sea). Asia includes the world's most populous country (China) and the world's least densely populated country (Mongolia). The continent includes as well some of the world's poorest countries (such as Bangladesh) and some of the world's wealthiest countries (such as Japan).

This chapter focuses on the southeastern quarter of the vast Eurasian land mass, covering a broad swath of countries that runs from Pakistan in the west to Japan in the east. Geographers use the term **Monsoon Asia** to refer to this portion of the continent because of the significant role played by the seasonal shift of wind patterns throughout the entire area. The countries located in the Caucasus Region and Central Asia were reviewed in Region 2; the countries of Southwest Asia and the Arabian Peninsula were covered in Region 3.

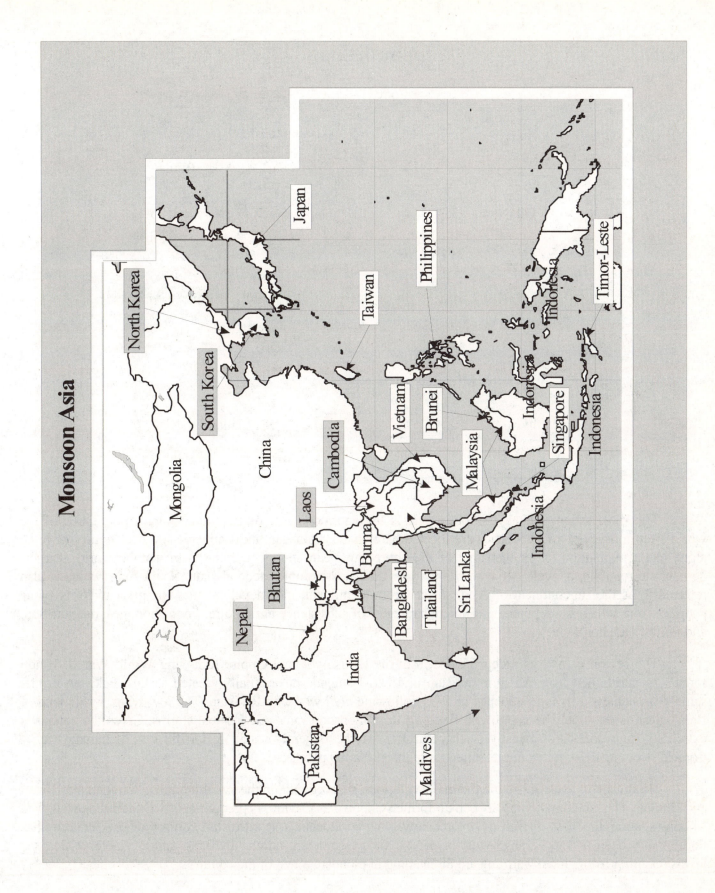

Monsoon Asia

Japan

Philippines

North Korea

Taiwan

South Korea

Timor-Leste

Indonesia

Vietnam

Brunei

Indonesia

China

Mongolia

Cambodia

Malaysia

Singapore

Indonesia

Laos

Burma

Indonesia

Bhutan

Bangladesh

Thailand

Sri Lanka

Nepal

India

Pakistan

Maldives

South Asia

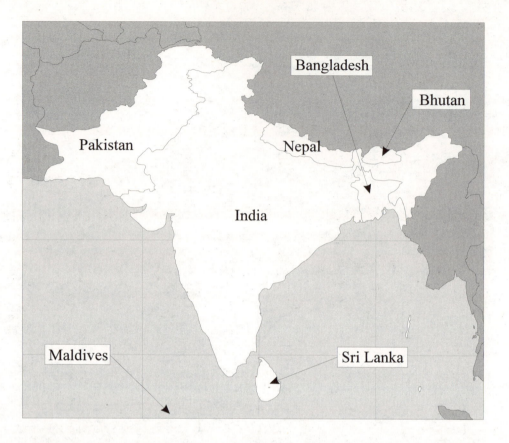

The most salient feature of South Asia is the enormous size of its human population. Three of the world's ten most populous countries are located here (India, Pakistan, and Bangladesh). The country of India alone has a population larger than that of the entire European continent. India is by far the world's largest democratic republic, as well as the world's largest and most complex federal state. South Asia, which is also referred to as the subcontinent of Asia, suffers from significant underdevelopment. The great majority of its people live in villages and pursue inefficient strategies of subsistence agriculture. Food shortages, malnutrition, and famines are common.

The region gained its independence from the British colonial empire following World War II. Hindu India was partitioned from Muslim Pakistan in 1947. Bangladesh, originally created as East Pakistan in that same year, achieved its independence in 1971 following civil war with Pakistan. The low-lying, fertile land of Bangladesh is extremely susceptible to flooding from monsoon rains and tropical storms. Given the country's extremely high population density, such natural disasters frequently result in horrific loss of human life. A tropical storm in 1970, for example, killed more than 300,000 people.

Hinduism and Buddhism are the major religions, respectively, in the small mountain kingdoms of Nepal and Bhutan. The island nation of Sri Lanka, formerly known as Ceylon, is predominantly Buddhist as well. The Maldives, an archipelago of well over 1,200 low-lying coral islands, is a Muslim country whose economy has traditionally been tied to fishing. Most of the Maldive Islands are uninhabited and uninhabitable. Rising sea levels resulting from global warming are threatening to inundate much of the country.

PAKISTAN
(PAH-kih-stahn)

Population	164,742,000 *(485 per square mile)*
Language(s)	Urdu & English
Area	339,732 *square miles*
Capital	**Islamabad** (iss-LAHM-uh-bahd)
Economy	$2,600 *annual per capita GDP*
Sovereignty	Federal Republic *independent since* 1947

INDIA

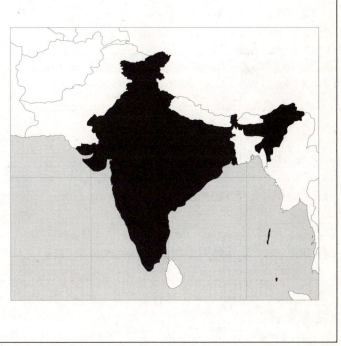

Population	1,129,866,000 *(924 per square mile)*
Language(s)	Hindi & English plus Hindustani and 14 other official languages
Area	1,222,510 *square miles*
Capital	**New Delhi**
Economy	$3,800 *annual per capita GDP*
Sovereignty	Federal Democratic Republic *independent since* 1947

NEPAL
(nuh-PAWL)

Population	28,902,000 *(509 per square mile)*
Language(s)	Nepali
Area	56,827 *square miles*
Capital	**Kathmandu** (kat-man-DOO)
Economy	$1,500 *annual per capita GDP*
Sovereignty	Democratic Republic *independent since 1768*

BHUTAN
(boo-TAHN)

Population	2,328,000 *(130 per square mile)*
Language(s)	Dzongkha
Area	17,954 *square miles*
Capital	**Thimphu** (THIM-poo)
Economy	$1,400 *annual per capita GDP*
Sovereignty	Monarchy *independent since 1949*

BANGLADESH

Population	150,448,000 *(2,706 per square mile)*
Language(s)	Bengali & English
Area	55,598 *square miles*
Capital	**Dhaka** (DAHK-uh)
Economy	$2,300 *annual per capita GDP*
Sovereignty	Parliamentary Democracy *independent since* 1971

SRI LANKA
(sree-lahn-kuh)

Population	20,926,000 *(826 per square mile)*
Language(s)	Sinhala & Tamil
Area	25,332 *square miles*
Capital	**Colombo** (kuh-LUM-boh) *(Sri Jayewardenepura Kotte is the legislative capital)*
Economy	$4,700 *annual per capita GDP*
Sovereignty	Republic *independent since* 1948

MALDIVES
(MAHL-deevz)

Population	369,000 *(3,209 per square mile)*
Language(s)	Dhivehi
Area	115 *square miles*
Capital	**Male** (MAH-lay)
Economy	$3,900 *annual per capita GDP*
Sovereignty	Republic *independent since* 1965

Southeast Asia

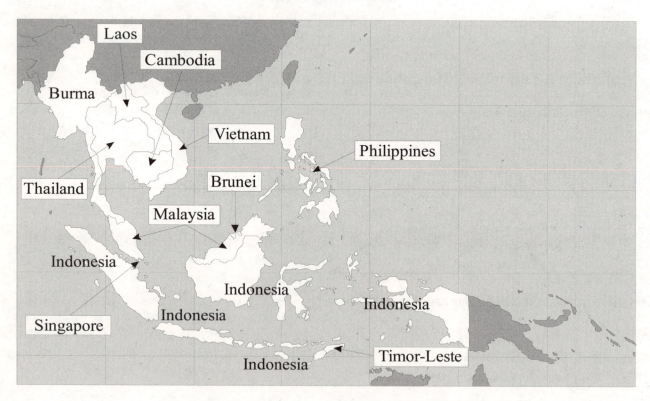

Southeast Asia is a predominantly tropical region consisting of a series of peninsulas and islands; only one of its countries, Laos, is landlocked. It is a region of wide ethnic and cultural diversity, especially given its relatively small geographic size. The region as a whole is much less densely populated than either South Asia or East Asia, but Southeast Asia does include one of the most populous countries on earth, Indonesia, which is larger than all but three of the world's 194 sovereign states.

The countries of mainland Southeast Asia have predominantly agricultural economies, despite the presence of large but unexploited mineral deposits, especially in Burma (also known as Myanmar) and Thailand (historically known as Siam). The source of more than half the world's opium crop can be found in the "Golden Triangle" at the border juncture between Burma, Laos, and Thailand. The republic of Cambodia was known as Kampuchea from 1976 to 1989, but its official name has since reverted to Cambodia. The socialist republic of Vietnam dates to 1975, when North and South Vietnam were united following the North Vietnamese victory in the Vietnam War (which the Vietnamese refer to as the American War). The former capital of South Vietnam, Saigon, has been renamed Ho Chi Minh City.

The insular countries of Southeast Asia include one of the world's richest states (Brunei) and some of the world's fastest growing and most prolific economies (Singapore and Malaysia). The city state of Singapore, located on a small group of islands at the tip of the Malay Peninsula, is one of the most densely populated countries in the world, and it is also a thriving commercial center for thousands of multinational corporations. Malaysia, which is split between the southern portion of the Malay Peninsula and the northern coast of the island of Borneo, is also rapidly industrializing.

Indonesia is the world's largest Muslim country. An archipelago of more than 13,500 islands, it stretches across 3,000 miles of the equatorial Pacific and includes the islands of Java, Sumatra, and Celebes as well as portions of the islands of Borneo and New Guinea. With more than 7,000 islands, the Philippines is the world's second largest archipelago. Roman Catholicism is the principal religion, making the Philippines the only Christian country in eastern Asia.

The Indonesian archipelago is also home to one of the world's newest countries, Timor-Leste, which became independent in 2002. Timor-Leste was a Portuguese possession until 1975 (unlike the rest of Indonesia, which had been a Dutch colony). When Portugal relinquished control of Timor-Leste in 1975, a civil war erupted between the factions that favored independence and those that favored integration with Indonesia. Indonesia intervened militarily, and subsequently integrated Timor-Leste as its 27th province (although the United Nations never recognized Indonesia's absorption of Timor-Leste). In an August 1999 referendum, the people of Timor-Leste voted overwhelmingly for independence, and pro-integration militias responded with a campaign of violence, looting, and arson; many people were killed, and more than 85% of all buildings were destroyed, including virtually all of the schools and businesses. The people of Timor-Leste elected a Constituent Assembly in 2001 which drafted and adopted a national constitution creating a democratic republic with guaranteed protections for individual liberties. Timor-Leste's independence was officially recognized by the United Nations on May 20, 2002. The new country of Timor-Leste has thirteen administrative districts, including the non-contiguous district of Oecussi (also called Ambeno) on the north coast of West Timor. Despite its extremely low per capita GDP, Timor-Leste's economic prospects are relatively bright. There are substantial oil and gas reserves in the nearby waters that have significantly increased government revenues in recent years, although the industry has done little to help the country's unemployment problem, because the high-technology production facilities are in Australia.

BURMA

(*also known as* Myanmar [mee-ahn-MAHR])

Population	47,374,000 *(181 per square mile)*
Language(s)	Burmese
Area	261,228 *square miles*
Capital	**Rangoon** (*the military regime refers to the capital as* Yangon) [*the administrative capital is* Naypyidaw]
Economy	$1,800 *annual per capita GDP*
Sovereignty	Republic (military regime) *independent since* 1948

THAILAND

(TIGH-land)

Population	65,068,000 *(328 per square mile)*
Language(s)	Thai & English
Area	198,115 *square miles*
Capital	**Bangkok**
Economy	$9,200 *annual per capita GDP*
Sovereignty	Constitutional Monarchy *independent since traditional founding date of* 1238 *(never colonized)*

LAOS
(LAUSS)

Population	6,522,000 *(71 per square mile)*
Language(s)	Lao, French & English
Area	91,429 *square miles*
Capital	**Vientiane** (vyen-TYAWN)
Economy	$2,200 *annual per capita GDP*
Sovereignty	Socialist Republic *independent since* 1949

CAMBODIA
(kam-BOH-dee-uh)

Population	13,996,000 *(200 per square mile)*
Language(s)	Khmer, French & English
Area	69,898 *square miles*
Capital	**Phnom Penh** (puh-NOM PEN)
Economy	$2,800 *annual per capita GDP*
Sovereignty	Constitutional Monarchy *independent since* 1953

VIETNAM
(vee-eht-NAHM)

Population	85,262,000 *(666 per square mile)*
Language(s)	Vietnamese, English & French
Area	128,066 *square miles*
Capital	**Hanoi** (han-NOY)
Economy	$3,100 *annual per capita GDP*
Sovereignty	Socialist Republic *independent since 1945*

MALAYSIA

Population	24,821,000 *(195 per square mile)*
Language(s)	Malay (*i.e.*, Bahasa Malaysia) *plus* English, Chinese, *& several others*
Area	127,320 *square miles*
Capital	**Kuala Lumpur** (KWAH-luh LOOM-poor)
Economy	$12,800 *annual per capita GDP*
Sovereignty	Constitutional Monarchy *independent since 1957*

SINGAPORE

Population	4,553,000 *(17,246 per square mile)*
Language(s)	Chinese, Malay, Tamil, & English
Area	264 *square miles*
Capital	**Singapore**
Economy	$31,400 *annual per capita GDP*
Sovereignty	Parliamentary Republic *independent since 1965*

BRUNEI
(BROO-nigh)

Population	375,000 *(168 per square mile)*
Language(s)	Malay, English & Chinese
Area	2,226 *square miles*
Capital	**Bandar Seri Begawan** (BUN-dahr SERRY buh-GAH-wun)
Economy	$25,600 *annual per capita GDP*
Sovereignty	Constitutional Monarchy *independent since 1984*

INDONESIA

Population	234,694,000 *(319 per square mile)*	
Language(s)	Bahasa Indonesia *(form of* Malay), *plus* English, Dutch, & Javanese	
Area	735,310 *square miles*	
Capital	**Jakarta** (jah-KAHR-tah)	
Economy	$3,900 *annual per capita GDP*	
Sovereignty	Republic *independent since* 1945	

TIMOR-LESTE
(TEE-more LEHSH-tay)

Population	1,085,000 *(189 per square mile)*	
Language(s)	Tetum & Portuguese	
Area	5,743 *square miles*	
Capital	**Dili**	
Economy	$800 *annual per capita GDP*	
Sovereignty	Republic *independence recognized since* 2002	

PHILIPPINES

Population	91,077,000 *(786 per square mile)*
Language(s)	Filipino & English
Area	115,831 *square miles*
Capital	**Manila**
Economy	$5,000 *annual per capita GDP*
Sovereignty	Republic *independent since* 1898

East Asia

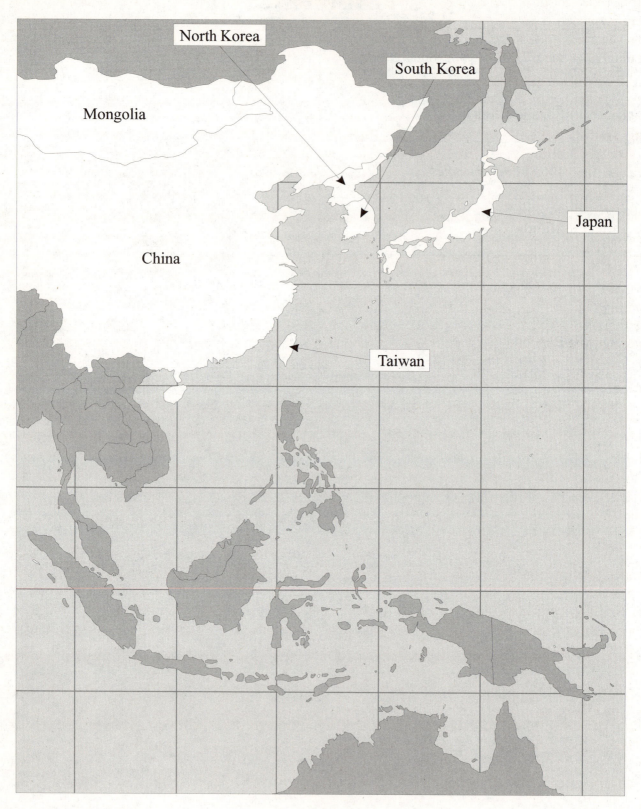

North Korea

South Korea

Mongolia

China

Japan

Taiwan

East Asia is dominated by the enormous country of China, which is overwhelmingly the most populous country in the world; one of every five people on earth is a citizen of China. China is bordered by the least densely populated country in the world, Mongolia, which serves as a buffer between China and Russia; the expansive deserts and steppes of Mongolia are inhabited chiefly by pastoral nomads, descendants of the Mongol hordes that once conquered much of Asia and Europe.

China emerged in its present political form following the Communist Revolution of 1949. Under the dictatorial leadership of Mao Zedong, the communists adopted economic policies that had disastrous effects, causing famines that cost the lives of tens of millions of people; the communists also imposed strict controls over everyday life that severely restricted political freedoms. Since 1978, Mao's successors have gradually introduced market-oriented reforms and decentralized economic decision making. The results have been remarkable: economic productivity has more than quadrupled during the past generation, and China now has the world's second largest GDP (although its *per capita* GDP remains relatively low). China recently surpassed Japan as the U.S.A.'s second-largest trading partner (behind Canada). Political control is still tight in China, even while economic controls are continuing to loosen.

Since 1949, there have in effect been two governments in China: the communist government, which is in control of the vast mainland, and the nationalist government, which is in control of the island of Taiwan (also called Formosa). Both governments have claimed to be the sole legitimate government of all of China (although Taiwan no longer maintains that claim), and both continue to maintain that China and Taiwan are integrally connected. (Indeed, some sort of political reintegration is likely to occur eventually.)

In terms of diplomatic recognition, Taiwan no longer exists as a sovereign state. Its seat in the United Nations was given to communist China in 1972, and the United States shifted its recognition to the mainland in 1978. (Although the U.S. no longer maintains diplomatic relations with Taiwan, it does maintain unofficial commercial and cultural relations with the society through the American Institute in Taiwan, which has offices in both the U.S. and Taiwan.) While not an officially sovereign state as far as most of the world is concerned, Taiwan is nevertheless an independent country by every practical measure: it has a constitutional government, a well-equipped military, and a thriving export economy.

Eventual reunification is also likely for North Korea and South Korea, especially now that the Cold War has ended. The peoples of the two countries are identical in all respects except recent history: they share the same culture, language, and ethnicity. Today, however, there is a major economic difference between the two. North Korea is rural and underdeveloped, while South Korea is urban and industrialized, with an export economy comparable to that of Taiwan and Singapore. The South Korean capital, Seoul, is one of the largest cities in the world.

The most prosperous country of East Asia, obviously, is the island nation of Japan. Although the country is comprised of more than 3,000 islands, 98 percent of the land area is contained in the four main islands of Honshu, Hokkaido, Kyushu, and Shikoku. Japan emerged from the devastation of World War II to build one of the world's largest and most efficient industrial economies. The country is a leading producer of many consumer and industrial products, and has one of the highest per capita GDP's in the world. Japan has maintained a consistent and sizeable trade surplus with the United States for more than two decades.

CHINA

Population	1,321,852,000 *(358 per square mile)*
Language(s)	Chinese
Area	3,690,045 *square miles*
Capital	**Beijing** (bay-ZHING)
Economy	$7,800 *annual per capita GDP*
Sovereignty	Socialist Republic *established* 1949, *but China is the world's oldest continuous civilization dating back more than 3,500 years*

TAIWAN
(tigh-WAHN)

Population	22,859,000 *(1,644 per square mile)*
Language(s)	Chinese
Area	13,901 *square miles*
Capital	**Taipei** (tigh-PAY)
Economy	$29,600 *annual per capita GDP*
Sovereignty	Democratic Republic *functionally independent since* 1949

MONGOLIA

Population	2,952,000 *(5 per square mile)*
Language(s)	Mongolian (*i.e.*, Khalkha Mongol)
Area	604,829 *square miles*
Capital	**Ulaanbaatar** (oo-LAHN BAH-tahr) (*also spelled* Ulan Bator)
Economy	$2,100 *annual per capita GDP*
Sovereignty	Republic *independent since* 1921

NORTH KOREA
(Democratic People's Republic of Korea)

Population	23,302,000 *(501 per square mile)*
Language(s)	Korean
Area	46,540 *square miles*
Capital	**Pyongyang** (pee-ong-YAHNG)
Economy	$1,800 *annual per capita GDP*
Sovereignty	Socialist Republic (dictatorship) *independent since* 1945

SOUTH KOREA
(Republic of Korea)

Population	49,045,000 *(1,283 per square mile)*
Language(s)	Korean
Area	38,230 *square miles*
Capital	**Seoul** (sohl)
Economy	$24,500 *annual per capita GDP*
Sovereignty	Republic *independent since 1945*

JAPAN

Population	127,433,000 *(874 per square mile)*
Language(s)	Japanese
Area	145,850 *square miles*
Capital	**Tokyo**
Economy	$33,100 *annual per capita GDP*
Sovereignty	Constitutional Monarchy *established 1947, but the origins of the Japanese state can be traced back at least 2,500 years (traditional founding 660 BCE)*

Review Exercises
for
Monsoon Asia

Population Language Area Capital Economy Sovereignty

Country & Capital Identification

1.   List the *capital city* of each of the 24 independent countries of Monsoon Asia, and then indicate the location of each country on the map on the opposite page.

Country	Capital
Bangladesh	
Bhutan	
Brunei	
Burma	
Cambodia	
China	
India	
Indonesia	
Japan	
Laos	
Malaysia	
Maldives	
Mongolia	
Nepal	
North Korea	
Pakistan	
Philippines	
South Korea	
Sri Lanka	
Taiwan	
Thailand	
Timor-Leste	
Vietnam	
Singapore	

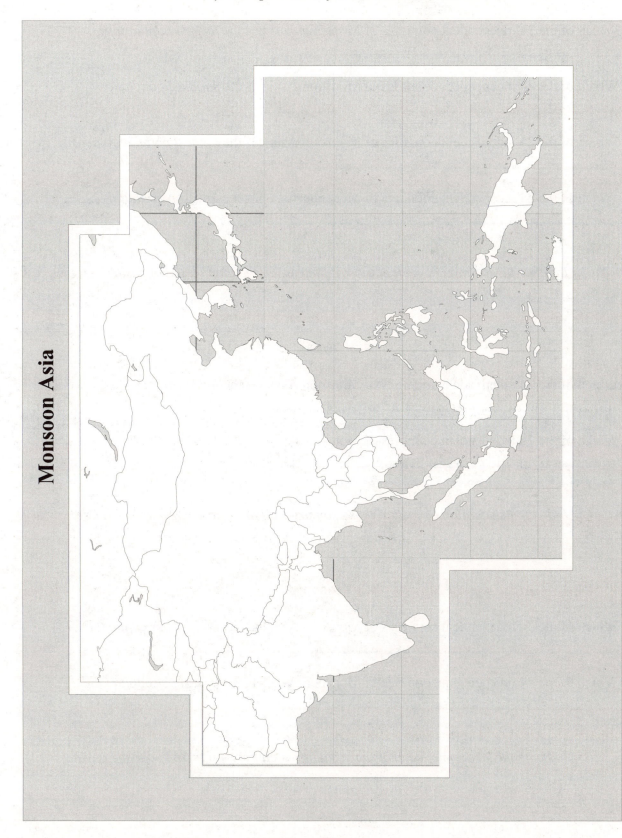

Monsoon Asia

Population

2.	Which of the 24 independent countries of Monsoon Asia has the *largest population*?

_____

3.	Which of the 24 independent countries of Monsoon Asia has the *smallest population*?

_____

4.	Which of the 24 independent countries of Monsoon Asia has the *highest population density*?

_____

5.	Which of the 24 independent countries of Monsoon Asia has the *lowest population density*?

_____

Indicate the location of each of those countries on the map on the opposite page.

Area

6.	Which of the 24 independent countries of Monsoon Asia has the *largest area*?

_____

7.	Which of the 24 independent countries of Monsoon Asia has the *smallest area*?

_____

Indicate the location of each of those countries on the map on the opposite page.

Economy

8.	Which of the 24 independent countries of Monsoon Asia has the *highest per capita GDP*?

_____

9.	Which of the 24 independent countries of Monsoon Asia has the *lowest per capita GDP*?

_____

Indicate the location of each of those countries on the map on the opposite page.

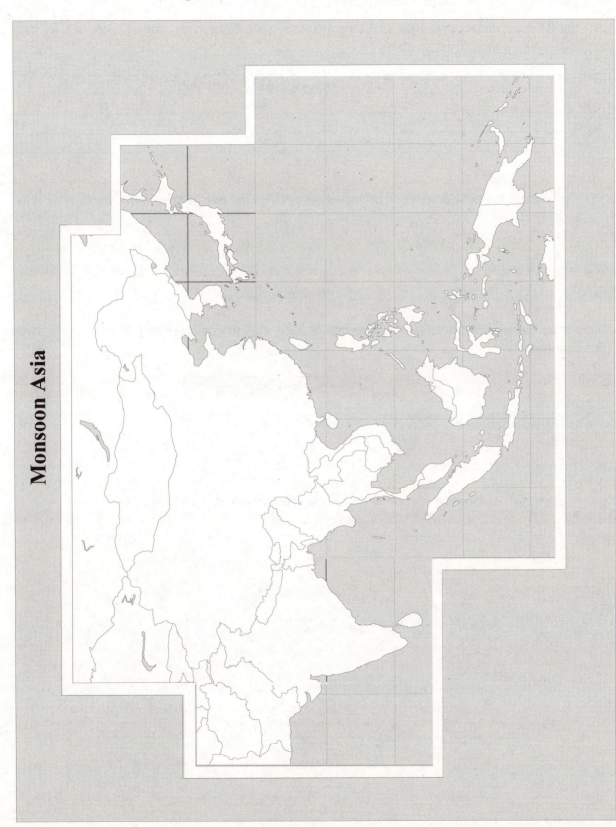

Monsoon Asia

Languages

10. 5 of the 24 independent countries of Monsoon Asia have *Chinese* as one of their official or principal languages. Name those five countries:

Five Chinese-Speaking Countries			
1		4	
2		5	
3			

Indicate the location of each of those countries on the map on the opposite page.

Sovereignty

11. There are *four socialist republics* among the 24 independent countries of Monsoon Asia. Name those four countries:

Four Socialist Republics			
1		3	
2		4	

Indicate the location of each of those countries on the map on the opposite page.

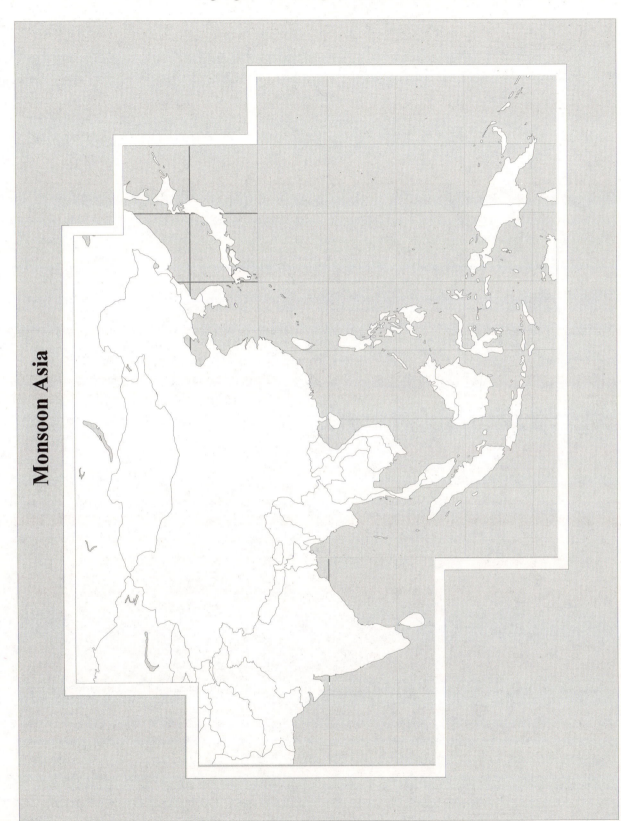

Monsoon Asia

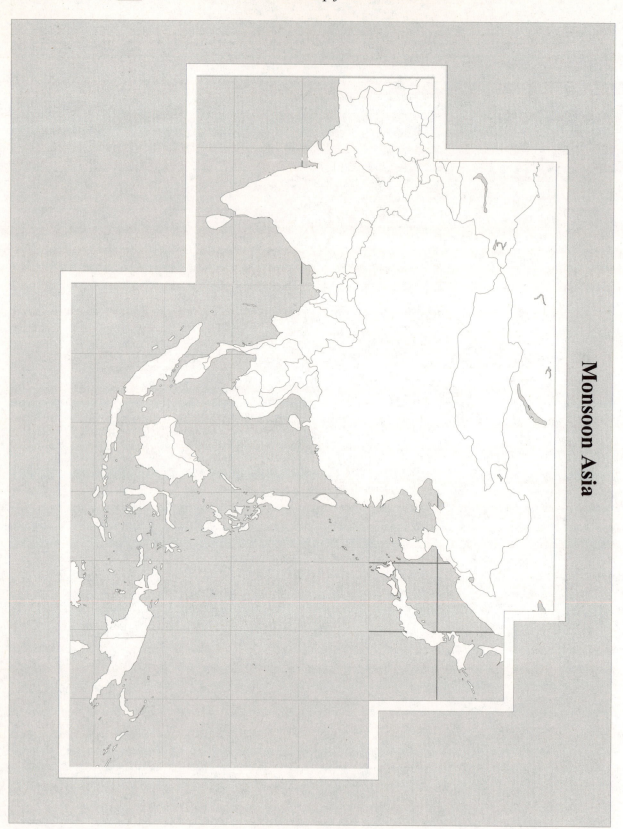

Monsoon Asia

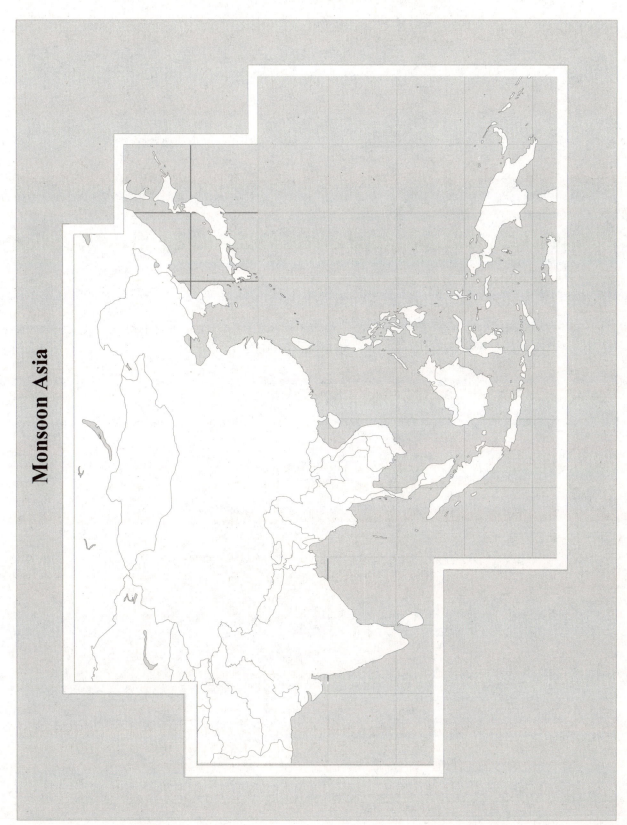

Monsoon Asia

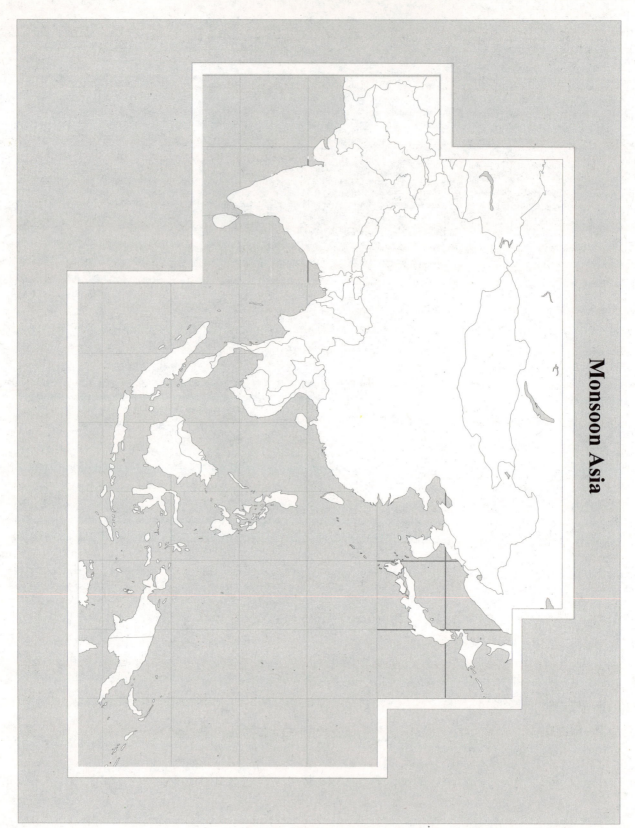

Monsoon Asia

Places of the World

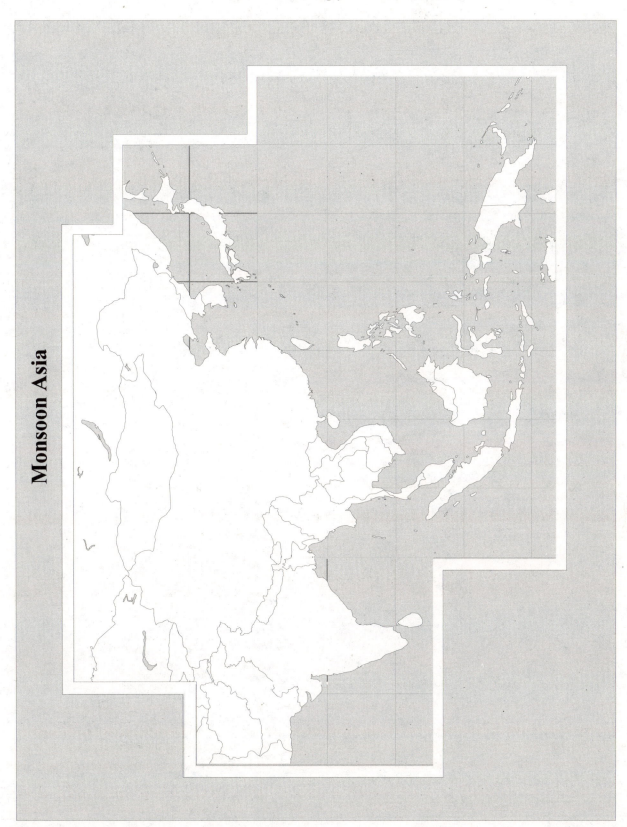

Monsoon Asia

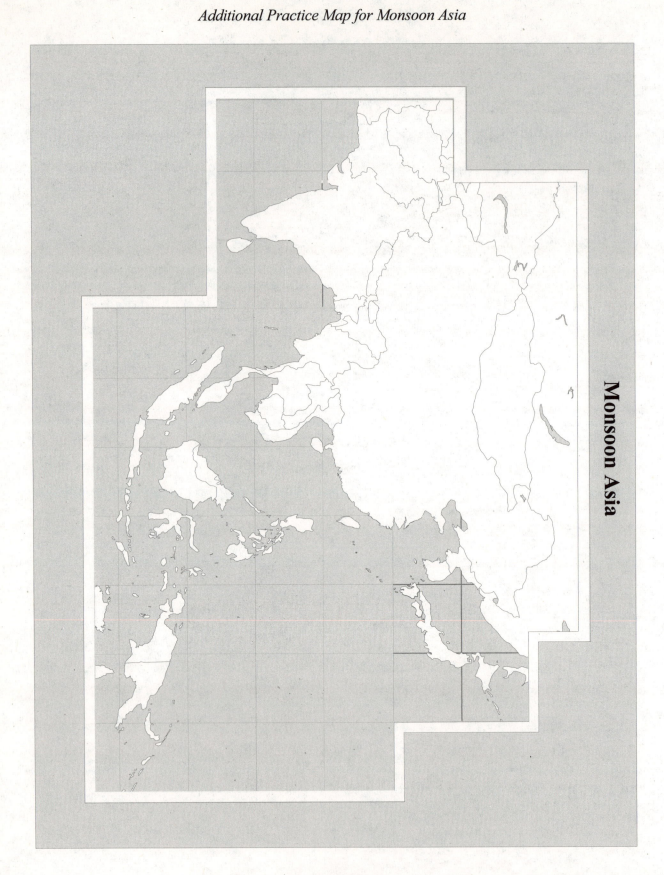

Monsoon Asia

REGION 5
Oceania

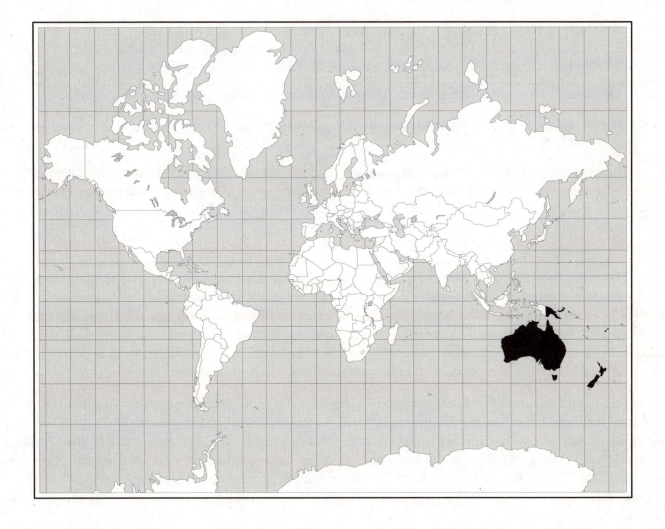

Oceania

There are 14 independent countries in Oceania (also known as the Pacific World):

Australia	**Micronesia**	**Papua New Guinea**	**Tuvalu**
Fiji	**Nauru**	**Solomon Islands**	**Samoa**
Kiribati	**New Zealand**	**Tonga**	**Vanuatu**
Marshall Islands	**Palau**		

The 14 independent countries of Oceania covered in this chapter can be conveniently divided into four groups: **Australia & New Zealand**, **Melanesia**, **Micronesia**, and **Polynesia**.

Australia & New Zealand	Australia, New Zealand
Melanesia	Papua New Guinea, Solomon Islands, Vanuatu, Fiji
Micronesia	Palau, Micronesia, Marshall Islands, Nauru
Polynesia	Kiribati, Tuvalu, Samoa, Tonga

The continent of **Australia** is located entirely within the Southern Hemisphere. Surrounded by the Pacific and Indian Oceans, it straddles the Tropic of Capricorn, and extends from the tropical waters of the Coral Sea to the temperate reaches of the Tasman Sea. Australia ranks 7th in size among the world's seven continents in terms of area, and 6th in size in terms of population (although it is the least populated of the inhabited continents).

The entire continent of Australia is occupied by a single independent country that bears the same name. The independent country of New Zealand is located well over 1,000 miles southeast of Australia. The region known as Oceania comprises a very large portion of the southwest Pacific Ocean and is traditionally subdivided into Melanesia, Micronesia, and Polynesia. ("Nesia" means "island;" the prefix in each case describes a characteristic feature of the subregion.) Oceania includes thousands of islands, but only a dozen sovereign states. The principal political dependencies of the region will be covered in *Appendix A*.

The Australian continent is unique in many respects. Besides being the smallest of the world's seven continental landforms, it is also the least mountainous. As the only inhabited continent that lacks a land bridge to other continental areas, Australia is the most remote continent in the world (excluding uninhabited Antarctica, of course). Australia's geographic isolation has had important consequences for its natural history. It is the only continent in the world, for example, where the majority of the indigenous mammals are either marsupials (pouched mammals, such as the kangaroo) or monotremes (egg-laying mammals, such as the duck-billed platypus).

At the same time, Australia would seem very familiar to people in the United States. Australia is the one country in the world that is most similar to the U.S.A. in terms of history, culture, language, political organization, and geographic size, among other features. Australia and the United States both began as British colonies; both were settled primarily by Europeans who displaced the native populations; both occupy vast territories with lengthy coastlines on two major oceans; and both, of course, are English-speaking democracies.

Oceania

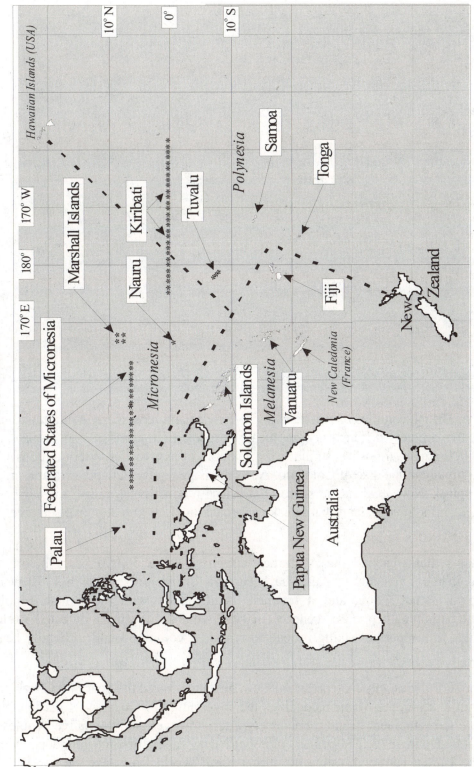

Australia and New Zealand

Australia

New Zealand

Tasmania (Australia)

    The former British colonies of Australia and New Zealand have many cultural and historical features in common, but the two are very different geographically, especially in terms of size, climate, and topography. At just under 3,000,000 square miles, Australia is comparable in size to the 48 contiguous United States, while New Zealand is only about the size of Colorado. Australia is generally flat and arid, and much of the country is tropical or subtropical. New Zealand, in sharp contrast, is mountainous, rainy, and temperate. The mountain range on New Zealand's South Island, with peaks approaching 12,000 feet above sea level, is aptly named the Southern Alps.

    Aboriginal settlers reached the Australian continent some 50,000 years before the first European explorers arrived in the 17th century. Australia became a British possession in 1770 when it was formally claimed by Captain James Cook, and it retained its affiliation with the British Empire after becoming independent at the beginning of the 20th century. In 1999, voters in Australia defeated a referendum to change the country's status to a republic, opting instead to remain a commonwealth officially headed by the British monarch.

    The Maori, a Polynesian people, reached New Zealand around the year 800 C.E. Under the terms of the Treaty of Waitangi in 1840, the Maori recognized the sovereignty of Queen Victoria while retaining territorial rights. British colonial settlement began immediately after, however, and the native peoples were defeated in a series of wars that ended in 1872. The British colony of New Zealand became an independent dominion in 1907 and supported the United Kingdom militarily during borth World Wars. By the 1980's, New Zealand's full participation in a number of defense alliances had lapsed. In recent years, the government of New Zealand has sought to address longstanding Maori grievances.

AUSTRALIA

Population	20,434,000 *(7 per square mile)*
Language(s)	English
Area	2,969,910 *square miles*
Capital	**Canberra** *(Australian Capital Territory)*
Economy	$33,300 *annual per capita GDP*
Sovereignty	Parliamentary Democracy *independent since* 1901

NEW ZEALAND

Population	4,116,000 *(39 per square mile)*
Language(s)	English & Maori
Area	104,454 *square miles*
Capital	**Wellington**
Economy	$26,200 *annual per capita GDP*
Sovereignty	Parliamentary Democracy *independent since* 1907

Melanesia

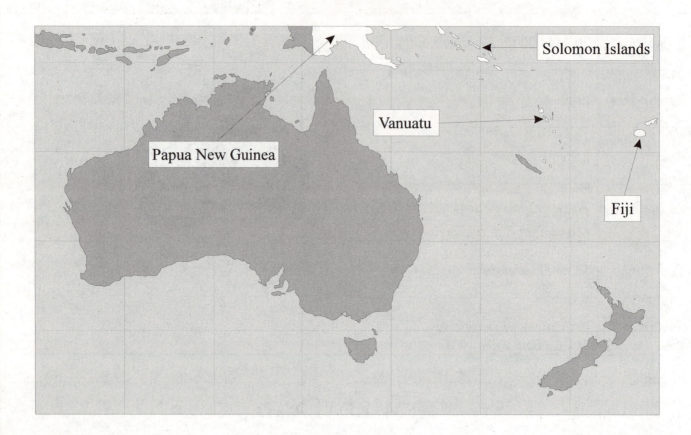

Melanesia is that portion of the Pacific realm immediately northeast of Australia that extends in a southeasterly arc from New Guinea through the Solomon Islands, Vanuatu, and Fiji (it includes as well the French dependency of New Caledonia, which will be covered in *Appendix A*). The indigenous inhabitants of Melanesia tend to have dark skin and dark hair, giving rise to the name of their region ("melas" means black). Papua New Guinea, which covers the eastern half of the island of New Guinea, is the largest independent country in Oceania after Australia, both in terms of geographic and demographic size. (Notice that the country includes the Bismarck Archipelago immediately off its northeast coast.) The mountainous interior of Papua New Guinea, which for centuries was inhabited by warlike tribal horticulturists, was not penetrated by outsiders until the 1930's. As a result of that long isolation, Papua New Guinea has the greatest linguistic diversity of any country on earth; some 700 languages are still identifiable on the island.

The Solomons are comprised of a twin chain of tropical islands extending almost 900 miles east-southeast of Papua New Guinea. Guadalcanal, site of an important World War II battle, is the largest of the Solomon Islands and the location of its capital city. Vanuatu, whose lyrical name means "Our Land Forever," consists of an 800-mile-long archipelago of about 80 islands. Vanuatu was formerly known as the New Hebrides. Fiji is a volcanic archipelago of some 330 islands dispersed over an area of approximately 250,000 square miles. Only about one-third of the islands are inhabited. Nearly four-fifths of the population lives on the largest island, Viti Levu, which is the site of the capital city.

PAPUA NEW GUINEA
(PAHP-yoo-uh new GIHN-ee)

Population	5,796,000 *(32 per square mile)*
Language(s)	English, Tok Pisin, & Hari-Motu
Area	178,704 *square miles*
Capital	**Port Moresby** (port MORZ-bee)
Economy	$2,700 *annual per capita GDP*
Sovereignty	Parliamentary Democracy *independent since* 1975

SOLOMON ISLANDS

Population	567,000 *(52 per square mile)*
Language(s)	English
Area	10,954 *square miles*
Capital	**Honiara**
Economy	$600 *annual per capita GDP*
Sovereignty	Parliamentary Democracy *independent since* 1978

VANUATU
(van-oo-AH-too)

Population	212,000 *(45 per square mile)*
Language(s)	English, French, & Bislama
Area	4,707 *square miles*
Capital	**Port Vila** *(also known simply as* Vila*)*
Economy	$2,900 *annual per capita GDP*
Sovereignty	Democratic Republic *independent since* 1980

FIJI
([FEE-jee] *also known as* Fiji Islands)

Population	919,000 *(130 per square mile)*
Language(s)	English, Fijian, & Hindustani
Area	7,056 *square miles*
Capital	**Suva** (SOO-vuh)
Economy	$6,200 *annual per capita GDP*
Sovereignty	Republic *independent since* 1970

Micronesia

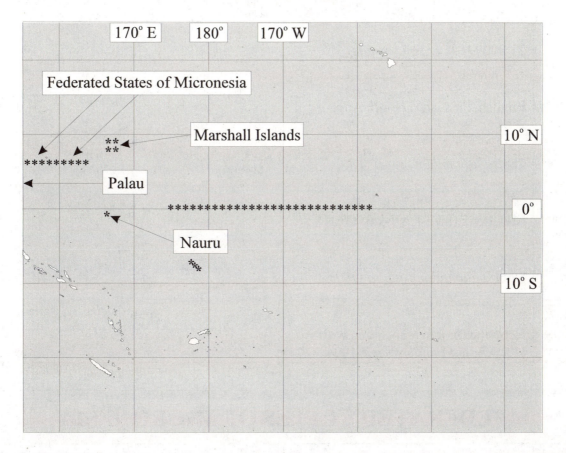

The region of Micronesia, which lies immediately north of Melanesia and extends from just east of the Philippines to the vicinity of the 180ᵗʰ meridian, is named for the tiny size of its 2,000 or so islands, the majority of which are low-lying coral atolls. The region includes four independent countries: Palau, Federated States of Micronesia, Marshall Islands, and Nauru. The first three are all sovereign states in "Free Association" with the United States (the U.S.A. retains rights for their defense but otherwise recognizes their autonomy). Nauru, the world's third smallest country in both area and population, is the only independent country in Oceania that consists of a single island. Located just south of the equator between the Marshall Islands and the Solomon Islands, Nauru at one time boasted one of the world's highest per capita incomes. The country's wealth was based on large deposits of high-grade phosphate accumulated from millennia of seabird droppings, but those deposits are now depleted, and the country is facing severe economic hardship.

Palau, a collection of more than 200 islands located at the western extremity of the Caroline Islands, is the world's fourth newest sovereign state after Serbia, Montenegro, and Timor-Leste. Also called Belau, this former Trust Territory of the United States achieved its independence on October 1, 1994. The Federated States of Micronesia encompasses the entire Caroline archipelago with the exception of Palau. The country, which includes more than 600 islands, extends in an east-west line for nearly 2,000 miles. A federal republic, Micronesia includes four constituent states, each with its own indigenous language: Yap, Chuuk, Pohnpei, and Kosrae. The 33 islands of the Marshall Islands are arranged in two parallel chains about 80 miles apart. The Marshalls include the former U.S. atomic testing sites of Eniwetok and Bikini.

PALAU
(puh-LAU)

Population	21,000 *(112 per square mile)*
Language(s)	English, Palauan, Tobi, & Angaur
Area	188 *square miles*
Capital	**Melekeok** (meh-LAY-kay-OAK)
Economy	$7,600 *annual per capita GDP*
Sovereignty	Republic (in free association with U.S.A.); *independent since 1994*

FEDERATED STATES OF MICRONESIA

Population	108,000 *(399 per square mile)*
Language(s)	English *plus several indigenous languages*
Area	271 *square miles*
Capital	**Palikir** (pah-lee-KIR)
Economy	$2,300 *annual per capita GDP*
Sovereignty	Republic (in free association with U.S.A.); *independent since 1986*

MARSHALL ISLANDS

Population	62,000 *(886 per square mile)*
Language(s)	English & Marshallese
Area	70 *square miles*
Capital	**Majuro**
Economy	$2,900 *annual per capita GDP*
Sovereignty	Republic (in free association with U.S.A.); *independent since 1986*

NAURU
(nah-OO-roo)

Population	14,000 *(1,750 per square mile)*
Language(s)	Nauruan & English
Area	8 *square miles*
Capital	**Yaren District** *(location of government offices— there is no "official" capital)*
Economy	$5,000 *annual per capita GDP*
Sovereignty	Republic *independent since 1968*

Polynesia

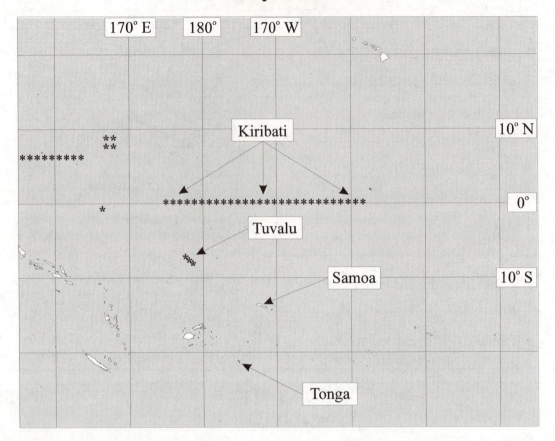

Polynesia ("many islands") comprises the largest area within Oceania, covering an enormous triangle whose three points are the Hawaiian Islands to the north, Easter Island to the southeast, and New Zealand to the southwest. Most of the thousands of islands scattered throughout the vastness of Polynesia are foreign possessions of European and American powers, particularly France, the United Kingdom, and the United States. The four independent countries in the region (Kiribati, Tuvalu, Samoa, and Tonga) include one, Kiribati, that actually overlaps the boundaries of Micronesia and Polynesia; the other three are all located on the western fringe of Polynesia, near the eastern boundaries of Melanesia and Micronesia.

Kiribati (notice that the spelling and pronunciation are not conventional) consists of three principal groups of islands scattered across 2 million square miles of the equatorial Pacific. The Gilbert Islands, which straddle the equator east of Nauru, are the westernmost group; the Phoenix Islands, south of the equator, are in the middle; and the Line Islands, which extend north of the equator, make up the easternmost portion of the country. (The Line Islands include Kiritimati, also known as Christmas Island.) Tuvalu, located north of Fiji, consists of nine small atolls totaling only 10 square miles, but those nine islands are scattered over some 500,000 square miles of the South Pacific. Tuvalu was formerly known as the Ellice Islands. Rising sea levels are threatening to inundate much of the country; since 2001, New Zealand has been accepting Tuvaluan immigrants as "environmental refugees." Samoa is part of the archipelago that includes the U.S. possession of American Samoa. The independent country of Samoa consists of two principal islands and a number of small islets. Tonga, southeast of Fiji, is also known as the Friendly Islands. It is made up of some 200 islands, although only about a quarter of them are inhabited.

KIRIBATI
(KIHR-ih-bahss)

Population	108,000 *(345 per square mile)*
Language(s)	English & I-Kiribati (Gilbertese)
Area	313 *square miles*
Capital	**Tarawa** (tuh-RAH-wuh) *[the administrative center is in* Bairiki *on the island of* Tarawa]
Economy	$2,800 *annual per capita GDP*
Sovereignty	Republic *independent since* 1979

TUVALU
(TOO-vuh-loo)

Population	12,000 *(1,200 per square mile)*
Language(s)	English & Tuvaluan
Area	10 *square miles*
Capital	**Funafuti** *[the administrative center is located at* Fongafale *on the island of* Funafuti]
Economy	$1,600 *annual per capita GDP*
Sovereignty	Parliamentary Democracy *independent since* 1978

SAMOA

(formerly known as Western Samoa)

Population	214,000 *(196 per square mile)*
Language(s)	Samoan & English
Area	1,093 *square miles*
Capital	**Apia** (AH-pee-uh)
Economy	$2,100 *annual per capita GDP*
Sovereignty	Parliamentary Democracy *independent since 1962*

TONGA

Population	117,000 *(466 per square mile)*
Language(s)	Tongan & English
Area	251 *square miles*
Capital	**Nuku'alofa** (NEW-koo-uh-LOH-fuh)
Economy	$2,200 *annual per capita GDP*
Sovereignty	Constitutional Monarchy *independent since 1970*

Review Exercises
for
Oceania

Population Language Area Capital Economy Sovereignty

Country & Capital Identification

1. List the *capital city* of each of the 14 independent countries of Oceania, and then indicate the location of each country on the map on the opposite page.

Country	Capital
Australia	
Fiji	
Kiribati	
Marshall Islands	
Micronesia	
Nauru	
New Zealand	
Palau	
Papua New Guinea	
Samoa	
Solomon Islands	
Tonga	
Tuvalu	
Vanuatu	

2. What is the literal meaning of the terms *Melanesia*, *Micronesia*, and *Polynesia*?

Melanesia:
_____

Micronesia:
_____

Polynesia:
_____

Indicate the location of these three areas on the map on the opposite page.

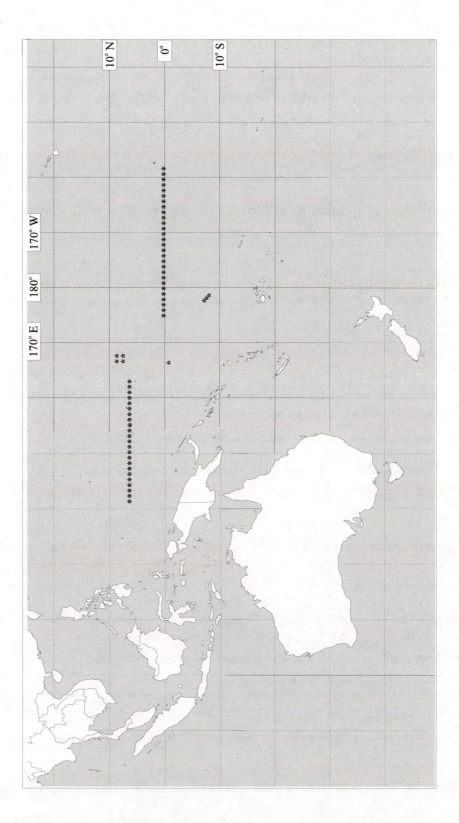

Population

3.   Australia has by far the largest population among the 14 independent countries of Oceania, but which country in the region has the *second largest population*?

_____

4.   Which of the 14 independent countries of Oceania has the *smallest population*?

_____

5.   Which of the 14 independent countries of Oceania has the *highest population density*?

_____

6.   Which of the 14 independent countries of Oceania has the *lowest population density*?

_____

Indicate the location of each of those countries on the map on the opposite page.

Area

7.   Australia has by far the largest land area among the 14 independent countries of Oceania, but which country in the region has the *second largest area*?

_____

8.   Which of the 14 independent countries of Oceania has the *smallest area*?

_____

Indicate the location of each of those countries on the map on the opposite page.

Economy

9.   Which of the 14 independent countries of Oceania has the *highest per capita GDP*?

_____

10.   Which of the 14 independent countries of Oceania has the *lowest per capita GDP*?

_____

Indicate the location of each of those countries on the map on the opposite page.

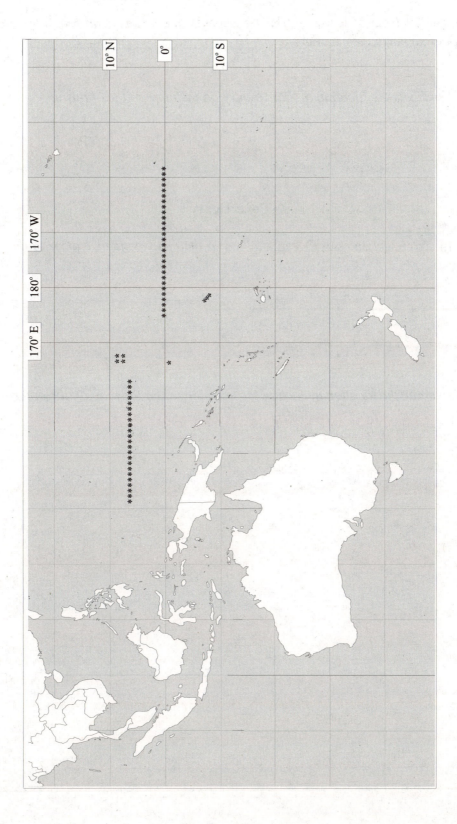

Languages

11.   Only one of the 14 independent countries of Oceania has a *European language other than English* as one of its principal languages, namely _____.

Indicate the location of that country on the map on the opposite page.

Sovereignty

12.   Two of the 14 sovereign states of Oceania *achieved their independence prior to the 1960's*. Name those two countries.

_____ &

_____

Indicate the location of each of those countries on the map on the opposite page.

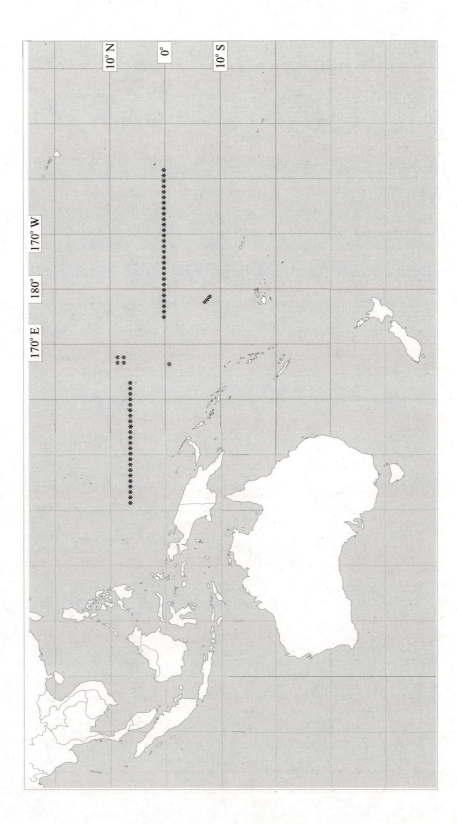

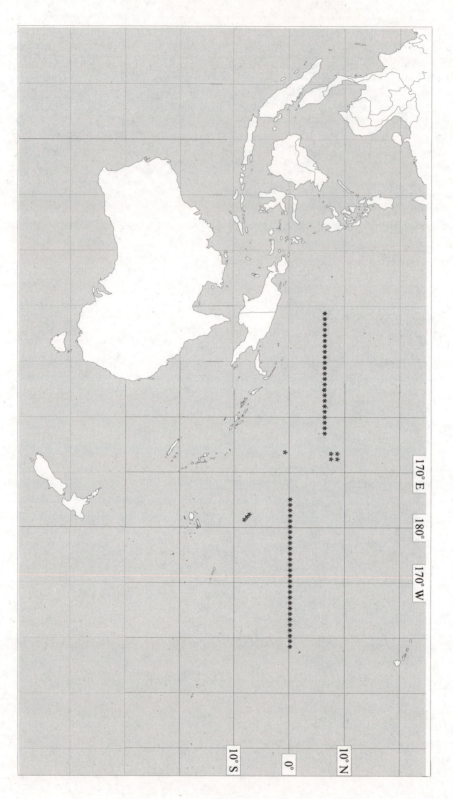

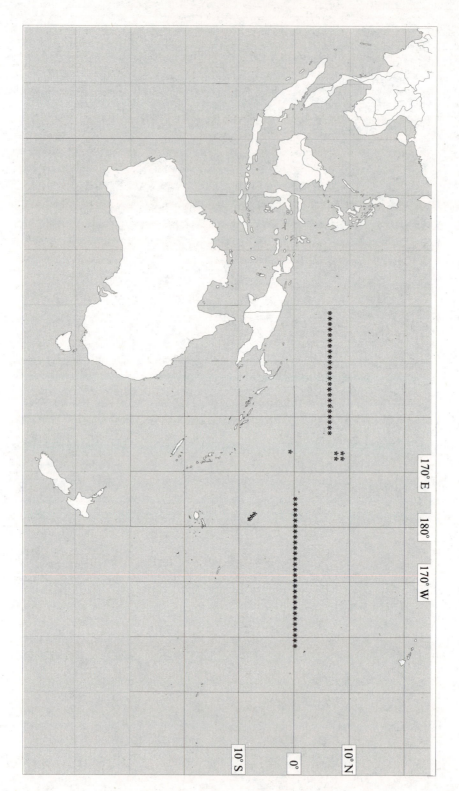

170° E

180°

170° W

10° S

0°

10° N

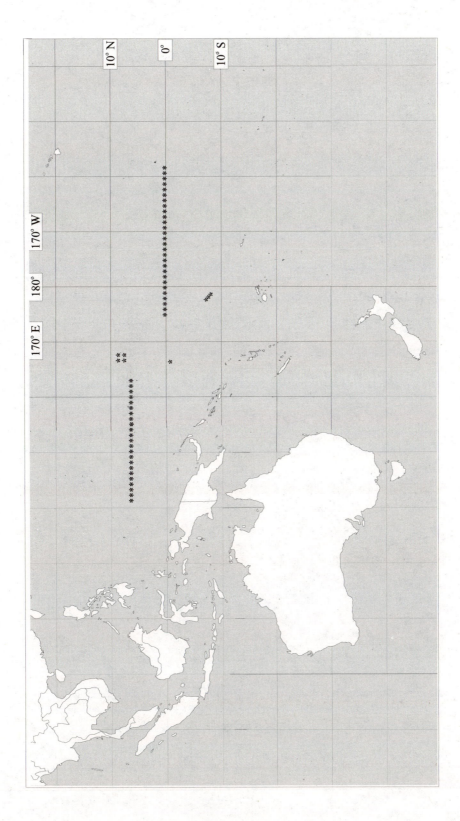

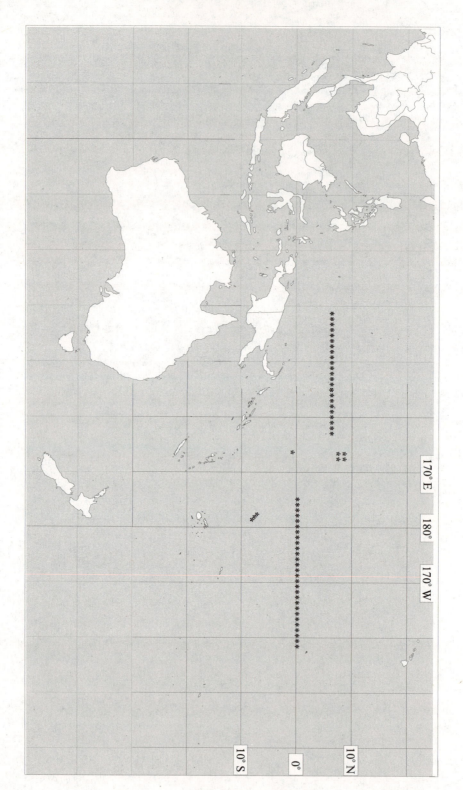

170° E

180°

170° W

10° S

0°

10° N

REGION 6
Sub-Saharan Africa

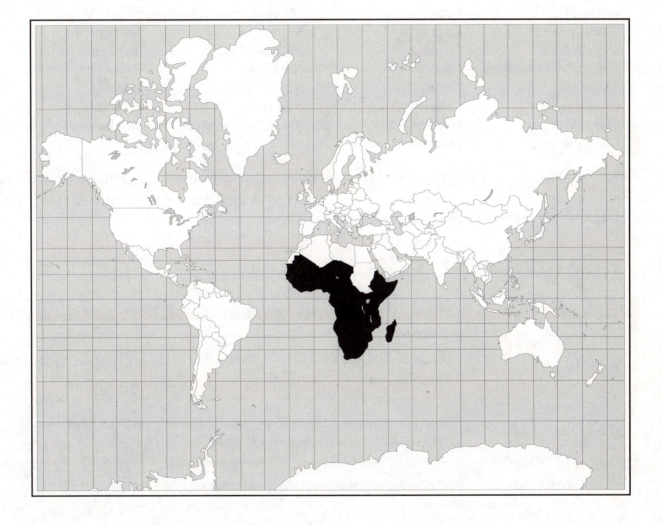

Sub-Saharan Africa

There are 47 independent countries in and around Sub-Saharan Africa:

Angola	**Côte d'Ivoire**	**Liberia**	**Senegal**
Benin	**Djibouti**	**Madagascar**	**Seychelles**
Botswana	**Equatorial Guinea**	**Malawi**	**Sierra Leone**
Burkina Faso	**Eritrea**	**Mali**	**Somalia**
Burundi	**Ethiopia**	**Mauritania**	**South Africa**
Cameroon	**Gabon**	**Mauritius**	**Swaziland**
Cape Verde	**Gambia**	**Mozambique**	**Tanzania**
Central African Rep.	**Ghana**	**Namibia**	**Togo**
Chad	**Guinea**	**Niger**	**Uganda**
Comoros	**Guinea-Bissau**	**Nigeria**	**Zambia**
Congo	**Kenya**	**Rwanda**	**Zimbabwe**
Congo Republic	**Lesotho**	**São Tomé & Príncipe**	

The 47 independent countries of Sub-Saharan Africa covered in this chapter can be conveniently divided into seven groups: the **Horn of Africa**, the **Sahel**, **West Africa**, **West Central Africa**, **East Africa**, **Southern Africa**, and **Indian Ocean Islands**.

The Horn of Africa	Eritrea, Ethiopia, Djibouti, Somalia
The Sahel	Mauritania, Mali, Burkina Faso, Niger, Chad, Senegal, Gambia, Cape Verde
West Africa	Guinea-Bissau, Guinea, Sierra Leone, Liberia, Côte d'Ivoire, Ghana, Togo, Benin, Nigeria
West Central Africa	Cameroon, Central African Republic, Equatorial Guinea, Gabon, Congo Republic, Congo, Sao Tomé & Príncipe
East Africa	Uganda, Kenya, Rwanda, Burundi, Tanzania
Southern Africa	Angola, Zambia, Malawi, Mozambique, Namibia, Botswana, Zimbabwe, South Africa, Lesotho, Swaziland
Indian Ocean Islands	Madagascar, Comoros, Mauritius, Seychelles

Sub-Saharan Africa

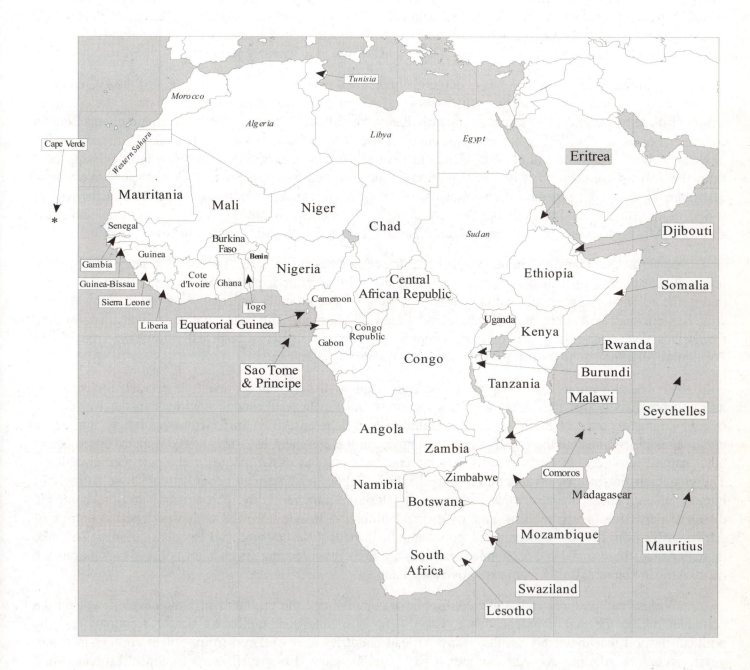

Tunisia

Morocco

Algeria

Libya

Egypt

Cape Verde

Western Sahara

Mauritania

Mali

Niger

Chad

Sudan

Eritrea

Djibouti

*

Senegal

Burkina
Faso

Nigeria

Central
African Republic

Ethiopia

Somalia

Gambia

Guinea

Benin

Cameroon

Guinea-Bissau

Cote
d'Ivoire

Ghana

Congo
Republic

Uganda

Kenya

Rwanda

Sierra Leone

Togo

Liberia

Equatorial Guinea

Gabon

Congo

Burundi

Tanzania

Sao Tome
& Principe

Malawi

Seychelles

Angola

Zambia

Comoros

Madagascar

Mauritius

Namibia

Zimbabwe

Botswana

Mozambique

South
Africa

Swaziland

Lesotho

Africa is the only continent to be crossed by both the Prime Meridian and the Equator. It extends from north of the 35th parallel in the Northern Hemisphere to south of the 30th parallel in the Southern Hemisphere, and stretches from west of the 15th meridian in the Western Hemisphere to east of the 50th meridian in the Eastern Hemisphere. Africa is connected to Asia by the Isthmus of Suez in Egypt (the Suez Canal forms the conventional boundary between the two continents). The African continent is bordered on the north by the Mediterranean Sea, on the west by the Atlantic Ocean, and on the east by the Red Sea and the Indian Ocean. Africa ranks 2nd in size among the world's seven continents in terms of area, and 2nd in size in terms of population.

The political map of Africa has undergone almost total revision since the beginning of the 20th century. In 1900, there were only two independent countries in Africa (Ethiopia and Liberia); the rest of the continent was controlled by various European powers, principally the British, French, German, Portuguese, and Belgian colonial empires. By the 1990's, however, the era of colonialism had come to a complete end. With the election of Nelson Mandela as President of South Africa in 1994, the last vestige of European minority rule disappeared from the African continent. Today, every portion of the continent is ruled by a native African government. There are 47 sovereign states on the continental mainland, and 6 independent island countries in the nearby waters of the Atlantic and Indian Oceans, for a total of 53 independent countries on the continent as a whole (the six North African countries of Morocco, Algeria, Tunisia, Libya, Egypt, and Sudan were covered in Region 3).

Nevertheless, many of the independent countries that emerged in Africa in the 20th century still bear the unmistakable imprint of their colonial past. Most of the political boundaries of today's countries were drawn during the colonial era, generally without regard for indigenous cultural boundaries. As a result, many African countries encompass a great deal of ethnic and linguistic diversity, and many native African peoples find themselves dispersed among two or more countries. European languages (especially English, French, and Portuguese) continue to be the official languages of a number of countries.

Viewed as a whole, there are two striking features about the African continent: its pervasive poverty and its rapid population growth. Africa is both the poorest and the fastest-growing continent in the world. The overwhelming majority of the 47 sovereign states of Sub-Saharan Africa suffer from seemingly intractable underdevelopment, despite the fact that many countries on the continent have substantial mineral deposits and other natural resources. In fact, Africa is the only continent that has grown poorer in the past quarter-century. The characteristic features of underdevelopment which plague most African countries constitute a litany of human misery. Those characteristics include high rates of infant mortality, chronic malnutrition, epidemic diseases, pervasive unemployment, and widespread illiteracy. In addition, the underdeveloped countries of Africa suffer from low life expectancies, poor sanitation, inadequate housing, and frequent famines. To make matters worse, these social and economic conditions are favorable breeding grounds for political corruption, and many African countries suffer from exploitive leadership.

Widespread political and ethnic conflict have compounded the problems of underdevelopment on the continent during the era of independence. Faced with desperate competition for scarce resources and the political turmoil that inevitably results, many African countries have suffered from violent conflict in recent years, including Algeria, Angola, Burundi, Chad, Congo, Cote d'Ivoire, Eritrea, Ethiopia, Liberia, Mali, Morocco, Namibia, Rwanda, Sierra Leone, Somalia, South Africa, Sudan, and Uganda. According to a recent study of civil wars around the world since 1960, there are three important risk factors for such violent conflicts: widespread poverty, low economic growth, and high dependence on natural resources. All three of those factors are very prevalent in Africa.

The Horn of Africa

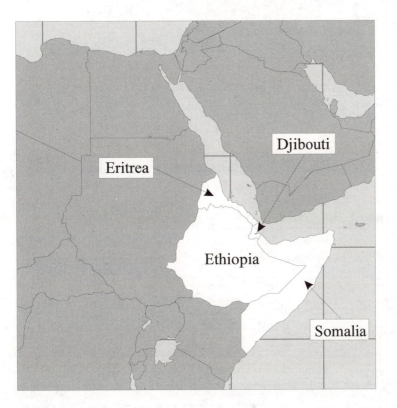

The Horn of Africa on the eastern side of the continent is named for the distinctive shape of its landform, which resembles a rhinoceros horn. The small country of Djibouti is strategically located at the entrance to the Red Sea. The region includes Africa's newest independent country, Eritrea, and the continent's oldest independent country, Ethiopia. A two-and-a-half-year border war between the two countries ended in 2000 under United Nations auspices, but despite the intervention of an international commission in 2002, Eritrea and Ethiopia have still not agreed on the final demarcation of their mutual border.

There is an area of disputed sovereignty in the Horn of Africa that concerns some 68,000 square miles of territory within the country of Somalia along the Gulf of Aden. In 1991, a rebel group calling itself the Somali National Movement announced its secession from Somalia, and declared the coastal region immediately east of Djibouti to be the independent Republic of Somaliland. Although neither the United Nations nor any foreign government has recognized its sovereignty, the entity has maintained a stable existence, aided by the overwhelming dominance of a ruling clan and the economic infrastructure left behind by British, Russian, and American military aid programs. The region bordering Somaliland has declared itself to be the autonomous state of Puntland, and it has been effectively self-governing since 1998; Puntland has territorial disputes with Somaliland concerning portions of their joint border. In 2000, representatives from several of the clans controlling various parts of the country formed a Transitional National Government with a three-year mandate to create a permanent and unified national government for Somalia; that mandate expired in 2003, however, without any progress towards a political solution. In late 2006, the Transitional Federal Government, backed by Ethiopian forces, took control of the capital, but it faces considerable political opposition and a multi-dimensional insurgency.

ERITREA
(air-ih-TRAY-uh *or* air-ih-TREE-uh)

Population	4,907,000 *(108 per square mile)*
Language(s)	Afar, Amharic, Arabic, Tigrinya, *& other indigenous languages*
Area	45,406 *square miles*
Capital	**Asmara** (*also spelled* Asmera)
Economy	$1,000 *annual per capita GDP*
Sovereignty	Republic *independent since* 1993

ETHIOPIA

Population	76,512,000 *(179 per square mile)*
Language(s)	Amharic, Tigrinya, English, *& several indigenous languages*
Area	426,373 *square miles*
Capital	**Addis Ababa** (ADD-iss AHB-uh-buh)
Economy	$1,000 *annual per capita GDP*
Sovereignty	Federal Republic, *consolidated in its modern form in 1889, but Ethiopia is the oldest independent country in Africa, with origins dating back 2,000 years*

214

DJIBOUTI
(jih-BOOT-ee)

Population	496,000 *(55 per square mile)*
Language(s)	French & Arabic
Area	8,958 *square miles*
Capital	**Djibouti**
Economy	$1,000 *annual per capita GDP*
Sovereignty	Republic *independent since* 1977

SOMALIA

Population	9,119,000 [*including* Somaliland] *(37 per square mile)*
Language(s)	Somali, Arabic, Italian, & English
Area	246,201 *square miles*
Capital	**Mogadishu**
Economy	$600 *annual per capita GDP*
Sovereignty	Republic *independent since* 1960

The Sahel

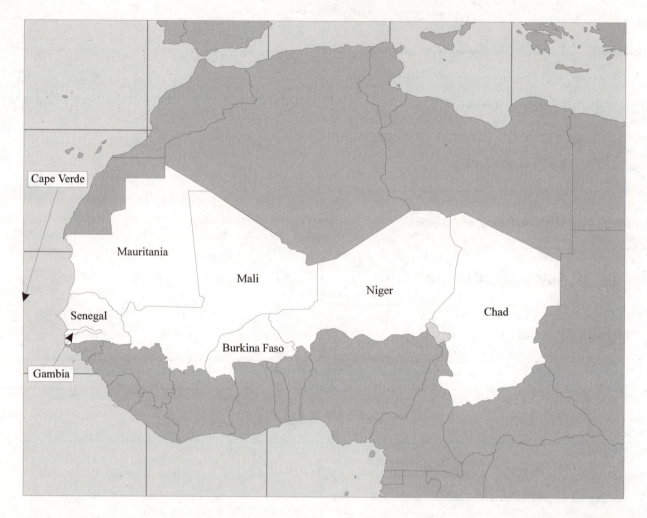

The Sahel (suh-HELL) is a broad belt of mixed tropical grasslands that runs along the southern border of the Sahara Desert. (*Sahel* is an Arabic word that means "coast" or "shore;" for the indigenous Arabs, the vast trackless expanse of the Sahara Desert is reminiscent of a great ocean.) The region, which extends from Mauritania on the Atlantic coast to Chad on the border with Sudan, is prone to severe droughts. Wide annual fluctuations in rainfall in the Sahel make life in the region especially precarious.

The region of the Sahel includes the insular country of Cape Verde, which is located in the Atlantic Ocean about 400 miles west of Senegal. A former Portuguese colony, Cape Verde is a small archipelago of ten islands and five islets. The majority of the country's population consists of people of mixed Portuguese and African ancestry. Mali and Senegal began their sovereign existence as the Mali Federation in 1960, when they gained their independence from France; Senegal withdrew from the federation after only a few months, however, and what had been called the Sudanese Republic was renamed Mali. Mauritania, Niger, and Chad also became independent from France in 1960, as did Burkina Faso (which was formerly known as Upper Volta). The former British possession of Gambia (also known as The Gambia) follows the course of the Gambia River, and is entirely embedded in the country of Senegal (although Gambia does have an Atlantic coastline at the mouth of the Gambia River).

216 Places of the World

MAURITANIA
(maw-ruh-TAY-nee-uh)

Population	3,270,000 *(8 per square mile)*
Language(s)	Arabic & Wolof *plus* French *and other indigenous languages*
Area	397,956 *square miles*
Capital	**Nouakchott** (NWOK-shot)
Economy	$2,600 *annual per capita GDP*
Sovereignty	Republic *independent since 1960*

MALI
(MAH-lee)

Population	11,995,000 *(25 per square mile)*
Language(s)	French & Bambara
Area	478,841 *square miles*
Capital	**Bamako** (BAHM-uh-koh)
Economy	$1,300 *annual per capita GDP*
Sovereignty	Republic *independent since 1960*

BURKINA FASO
(bur-KEEN-uh FAH-soh)

Population	14,326,000 *(135 per square mile)*
Language(s)	French, *plus indigenous languages in the Sudanic family*
Area	105,869 *square miles*
Capital	**Ouagadougou** (wah-guh-DOO-goo)
Economy	$1,400 *annual per capita GDP*
Sovereignty	Democratic Republic *independent since* 1960

NIGER
(nee-ZHAIR)

Population	12,895,000 *(26 per square mile)*
Language(s)	French, Hausa, Djerma, & Arabic
Area	489,192 *square miles*
Capital	**Niamey** (nee-AH-may)
Economy	$1,000 *annual per capita GDP*
Sovereignty	Republic *independent since* 1960

CHAD

Population	9,886,000 *(20 per square mile)*
Language(s)	French & Arabic *plus numerous indigenous languages*
Area	495,755 *square miles*
Capital	**N'Djamena** (inn-juh-may-nuh)
Economy	$1,500 *annual per capita GDP*
Sovereignty	Republic *independent since* 1960

SENEGAL

Population	12,522,000 *(165 per square mile)*
Language(s)	French & Wolof
Area	75,951 *square miles*
Capital	**Dakar** (duh-KAHR)
Economy	$1,800 *annual per capita GDP*
Sovereignty	Democratic Republic *independent since* 1960

GAMBIA

(*also known as* The Gambia)

Population	1,688,000 *(409 per square mile)*
Language(s)	English plus Mandinka, Wolof, Fula, *& others*
Area	4,127 *square miles*
Capital	**Banjul** (BAHN-jool)
Economy	$2,000 *annual per capita GDP*
Sovereignty	Democratic Republic *independent since 1965*

CAPE VERDE

Population	424,000 *(272 per square mile)*
Language(s)	Portuguese
Area	1,557 *square miles*
Capital	**Praia** (PRAH-yuh)
Economy	$6,000 *annual per capita GDP*
Sovereignty	Republic *independent since 1975*

West Africa

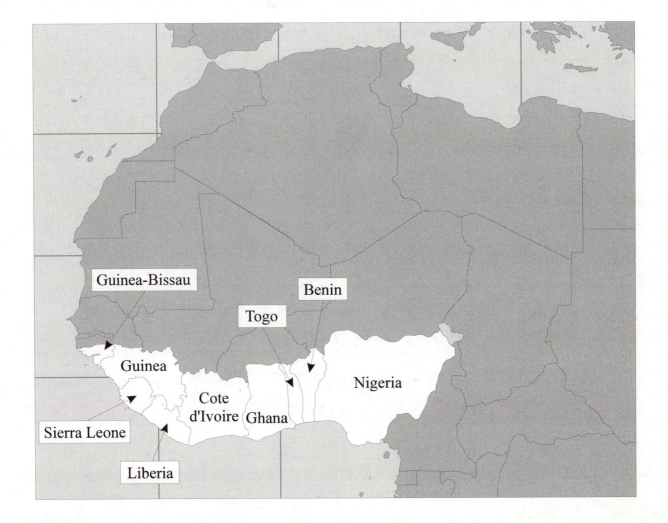

The French colonial empire, which maintained its capital at Dakar in Senegal, controlled most of the northwestern portion of the African continent (including, as we have seen, most of the countries in the Sahel). The French were unable to assert sovereignty over several areas along the coast, however, including Guinea-Bissau, which was controlled by the Portuguese, Liberia, which was never colonized, and Nigeria, Ghana, Sierra Leone, and Gambia, which constituted the scattered British possessions in the region.

Today the area is marked by considerable cultural diversity; there are hundreds of indigenous languages spoken in the region. Compared to the rest of the continent, West Africa has a relatively high population density. The region includes the continent's most populous country, Nigeria. Civil wars in Liberia and Sierra Leone have driven more than half-a-million refugees into Guinea.

A note about correct nomenclature: in 1985, the United Nations granted a request from the Ivorian government that "Côte d'Ivoire" be recognized as the sole official name of the country that had previously been known in English as the Ivory Coast.

GUINEA-BISSAU
(gihn-nee-bih-SAU)

Population	1,473,000 *(106 per square mile)*
Language(s)	Portuguese
Area	13,948 *square miles*
Capital	**Bissau**
Economy	$900 *annual per capita GDP*
Sovereignty	Republic *independent since 1974*

GUINEA
(GIHN-NEE [*also French pronunciation* gee-NAY])

Population	9,948,000 *(105 per square mile)*
Language(s)	French
Area	94,926 *square miles*
Capital	**Conakry** (KAHN-uh-kree)
Economy	$2,100 *annual per capita GDP*
Sovereignty	Republic *independent since 1958*

SIERRA LEONE

Population	6,145,000 *(222 per square mile)*
Language(s)	English, Mende, Temne, & Krio
Area	27,699 *square miles*
Capital	**Freetown**
Economy	$900 *annual per capita GDP*
Sovereignty	Democratic Republic *independent since* 1961

LIBERIA

Population	3,196,000 *(74 per square mile)*
Language(s)	English *plus many indigenous languages*
Area	43,000 *square miles*
Capital	**Monrovia**
Economy	$900 *annual per capita GDP*
Sovereignty	Republic *independent since* 1847

CÔTE d'IVOIRE

(KOHT deev-wahr)

Population	18,013,000 *(145 per square mile)*
Language(s)	French *plus many indigenous dialects*
Area	124,504 *square miles*
Capital	**Yamoussoukro** and **Abidjan** (ahb-ee-JAHN) (Yamoussoukro *has been the officially designated capital since 1983, but many government offices and most foreign embassies remain in* Abidjan)
Economy	$1,600 *annual per capita GDP*
Sovereignty	Republic, *independent since 1960*

GHANA

Population	22,931,000 *(249 per square mile)*
Language(s)	English *plus several indigenous languages*
Area	92,098 *square miles*
Capital	**Accra** (uh-KRAH)
Economy	$2,700 *annual per capita GDP*
Sovereignty	Democratic Republic *independent since 1957*

TOGO
(TOH-goh)

Population	5,702,000 *(260 per square mile)*
Language(s)	French, Ewe, Twi, Mina, Kabye, & Dagomba
Area	21,925 *square miles*
Capital	**Lomé** (loh-MAY)
Economy	$1,700 *annual per capita GDP*
Sovereignty	Emerging Democratic Republic *independent since* 1960

BENIN
(beh-NEEN)

Population	8,078,000 *(186 per square mile)*
Language(s)	French, Fon, Yoruba, *& several other indigenous languages*
Area	43,484 *square miles*
Capital	**Porto-Novo** *(Porto-Novo is the official capital—Cotonou is the seat of government)*
Economy	$1,100 *annual per capita GDP*
Sovereignty	Democratic Republic *independent since* 1960

NIGERIA

Population	135,031,000 *(379 per square mile)*
Language(s)	English, Hausa, Yoruba, Igbo, & Fulani
Area	356,669 *square miles*
Capital	**Abuja** (ah-BOO-jah)
Economy	$1,500 *annual per capita GDP*
Sovereignty	Republic *independent since* 1960

West Central Africa

West Central Africa is a region of dense equatorial rain forests and low population densities. It is an area of Africa that is relatively rich in natural resources. There are significant diamond, gold, copper, manganese, and uranium deposits throughout the region, for example, while Cameroon, Gabon, and Congo Republic all have marketable petroleum reserves. With the exception of Equatorial Guinea, which was controlled by the Spanish, and Congo, which constituted the Belgian Congo, the countries of Central Africa were all once part of French Equatorial Africa (although a portion of Cameroon was originally controlled by the Germans and subsequently by the British). The indigenous cultural boundaries overlap the contemporary political boundaries to a large extent. The Fang, for example, are distributed across Cameroon, Equatorial Guinea, and Gabon; the Bateke are found in Gabon and Congo Republic; and the Kongo make their home in Congo Republic, Congo, and Angola.

Notice that Equatorial Guinea consists of a mainland portion, known as Rio Muni, in addition to the island of Bioko (the country's high per capita GDP is derived primarily from offshore oil reserves). The small country of Sao Tomé and Principe is comprised of an archipelago of two main islands and four islets located in the Gulf of Guinea 125 miles west of Gabon. Like the other Atlantic Ocean African country, Cape Verde, Sao Tomé and Principe is a former Portuguese colony; also like its counterpart to the northwest, the majority of Sao Tomé's population is of mixed Portuguese and African ancestry. The former French colony of Ubangi-Shari became the Central African Republic when it achieved independence; the country has yet to achieve a stable and just political system. A small population, abundant natural resources, and considerable foreign support have helped make Gabon one of the more prosperous countries in Africa.

West Central Africa has recently been the scene of significant political turmoil. In 1997, the large country formerly known as Zaire changed its name to the Democratic Republic of Congo following a successful insurrection led by Laurent Désiré Kabila. Kabila's forces overthrew the corrupt government of President Mobutu Sese Seko, who had been in power for more than thirty years. (Mobutu Sese Seko had changed the country's name to Zaire in 1971; when he was removed in 1997, the name Zaire went with him.) Kabila failed to deliver the democratic government he had promised, however. In 1998, rebels backed by Rwanda and Uganda challenged Kabila's regime, and troops from Zimbabwe, Angola, Chad, and Sudan intervened to support the Kinshasa government. A cease-fire was signed in July 1999, but sporadic fighting continued. Laurent Kabila was assassinated in January 2001, and his son, Joseph Kabila, assumed power. In 2002 the younger Kabila was successful in getting occupying Rwandan forces to withdraw from eastern Congo; an agreement to end the fighting was signed shortly afterward by all warring parties, and Kabila was inaugurated as President in 2006.

The smaller country on the other side of the Congo River, formerly known as Congo, is now officially called the Republic of Congo, and variously referred to as "Congo Republic" or "Congo-Brazzaville." Meanwhile, the Democratic Republic of Congo (formerly Zaire) is now generally referred to simply as "Congo" or alternatively as "Congo-Kinshasa."

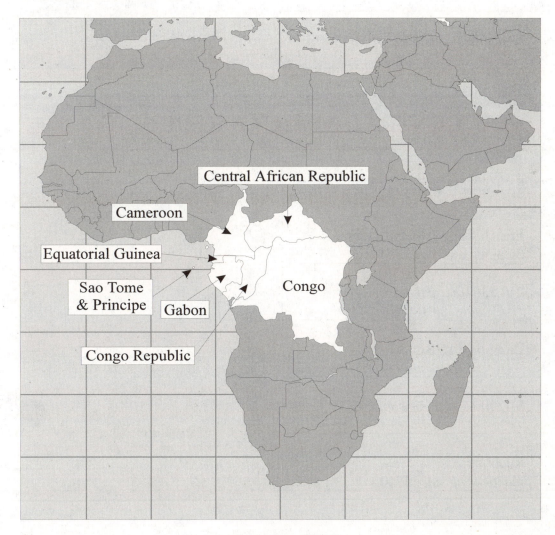

CAMEROON
(kam-uh-ROON)

Population	18,060,000 *(98 per square mile)*
Language(s)	French & English
Area	183,568 *square miles*
Capital	**Yaoundé** (yow-oon-DAY)
Economy	$2,500 *annual per capita GDP*
Sovereignty	Republic, *independent since* 1960

CENTRAL AFRICAN REPUBLIC

Population	4,369,000 *(18 per square mile)*
Language(s)	French, Sango, Arabic, Hunsa, & Swahili
Area	240,536 *square miles*
Capital	**Bangui** (BAHNG-ee)
Economy	$1,200 *annual per capita GDP*
Sovereignty	Republic *independent since* 1960

EQUATORIAL GUINEA

Population	551,000 *(51 per square mile)*
Language(s)	Spanish & French *plus* Fang, Bubi, & Ibo
Area	10,831 *square miles*
Capital	**Malabo** (mah-LAH-boh) [*formerly known as* Santa Isabel]
Economy	$50,200 *annual per capita GDP*
Sovereignty	Republic *independent since* 1968

GABON
(guh-BAHN)

Population	1,455,000 *(14 per square mile)*
Language(s)	French *plus several indigenous languages*
Area	103,347 *square miles*
Capital	**Libreville**
Economy	$7,100 *annual per capita GDP*
Sovereignty	Republic *independent since* 1960

REPUBLIC OF CONGO
(*former* "Congo" *now known as* "Congo Republic" *or* "Congo-Brazzaville")

Population	3,801,000 *(29 per square mile)*
Language(s)	French, Lingala, & Monokutuba *plus several other indigenous languages*
Area	132,047 *square miles*
Capital	**Brazzaville** (BRAHZ-uh-veel)
Economy	$1,400 *annual per capita GDP*
Sovereignty	Republic *independent since* 1960

DEMOCRATIC REPUBLIC OF CONGO
(*former* "Zaire" *now known as* "Congo" *or* "Congo-Kinshasa")

Population	65,752,000 *(73 per square mile)*
Language(s)	French, Swahili, & *several indigenous languages*
Area	905,446 *square miles*
Capital	**Kinshasa** (KEEN-shah-suh)
Economy	$700 *annual per capita GDP*
Sovereignty	Republic *independent since* 1960

SÃO TOMÉ AND PRÍNCIPE
(sau-too-may *and* pren-see-puh)

Population	200,000 *(538 per square mile)*
Language(s)	Portuguese
Area	372 *square miles*
Capital	**Sao Tomé**
Economy	$1,200 *annual per capita GDP*
Sovereignty	Republic *independent since* 1975

East Africa

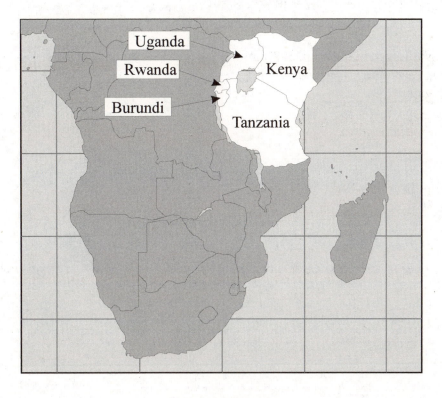

East Africa is a locale of enormous anthropological interest and immense significance for human evolution. The oldest human fossils have been found in the region centered on Kenya and Tanzania. It was in this area of the world, approximately five million years ago, where the first human ancestors began walking upright, and it was in this area of the world, approximately two-and-a-half million years ago, where the first representatives of the genus *Homo* appeared. East Africa was the stage upon which the first act of the drama of human existence was played out, and the geographic features of that stage have changed little in the past five million years. Now, as then, the region is dominated by vast dry savannas that support huge herds of herbivores.

East Africa includes the most densely populated country on the African mainland, Rwanda, which was the scene of astoundingly barbaric ethnic violence in 1994. Like its comparably-sized neighbor to the south, Burundi, Rwanda is tiny, overcrowded, landlocked, desperately poor, and ethnically diverse. Uganda, although somewhat larger both in area and population, shares essentially the same characteristics. Kenya is widely noted for its scenic attractions and its thriving tourist industry; the country has an expanding manufacturing sector as well. Notice that Tanzania includes the island of Zanzibar in the Indian Ocean (the name "Tanzania" was derived from the merger of mainland Tanganyika with insular Zanzibar).

UGANDA

Population	30,263,000 *(325 per square mile)*
Language(s)	English, Luganda, Swahili, & Arabic *plus several other indigenous languages*
Area	93,065 *square miles*
Capital	**Kampala** (kahm-PAH-luh)
Economy	$1,900 *annual per capita GDP*
Sovereignty	Republic *independent since* 1962

KENYA

Population	36,914,000 *(164 per square mile)*
Language(s)	English & Swahili *plus many other indigenous languages*
Area	224,961 *square miles*
Capital	**Nairobi** (nigh-ROH-bee)
Economy	$1,200 *annual per capita GDP*
Sovereignty	Republic *independent since* 1963

RWANDA
(roo-AHN-duh ["oo" as in "wood"])

Population	9,908,000 *(974 per square mile)*
Language(s)	Kinyarwanda, French, English, & Swahili
Area	10,169 *square miles*
Capital	**Kigali** (kih-GAH-lee)
Economy	$1,600 *annual per capita GDP*
Sovereignty	Republic *independent since* 1962

BURUNDI
(buh-ROON-dee ["oo" as in "wood"])

Population	8,391,000 *(781 per square mile)*
Language(s)	Kirundi, French, & Swahili
Area	10,745 *square miles*
Capital	**Bujumbura** (boo-jem-BOOR-uh) *["oo" as in "wood"]*
Economy	$700 *annual per capita GDP*
Sovereignty	Republic *independent since 1962*

TANZANIA
(tan-zuh-NEE-uh)

Population	39,384,000 *(108 per square mile)*
Language(s)	Swahili, English, Arabic, *and many other indigenous languages*
Area	364,900 *square miles*
Capital	**Dar es Salaam** (dahr ess suh-LAHM) *(the National Assembly meets in Dodoma [DOH-doh-mah], which is planned as a new national capital)*
Economy	$800 *annual per capita GDP*
Sovereignty	Republic *independent since 1964*

Southern Africa

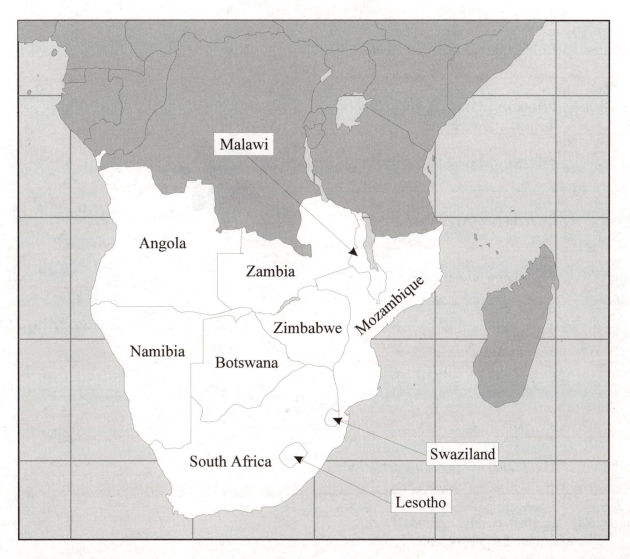

The region of Southern Africa is extraordinarily rich in natural resources, particularly in mineral reserves: gold, diamond, copper, platinum, iron, chromium, asbestos, lead, zinc, cobalt, and manganese deposits exist in great abundance. Nevertheless, the countries of the region are largely underdeveloped. Only small portions of the country of South Africa can be characterized as genuinely developed. Angola has yet to recover from the devastating effects of a lengthy civil war during the 1970's and 1980's. (Note that Angola includes the noncontiguous region of Cabinda, which is separated from the rest of the country by an arm of Congo.) Zambia (formerly Northern Rhodesia) is one of the world's largest copper producers, but it has suffered in recent years from falling copper prices and has accumulated a large foreign debt. Mozambique was a Portuguese colony for nearly five centuries until the third quarter of the 20th century; for a long time following independence, it was one of the world's poorest countries, and a majority of the population still lives in poverty. The recently independent country of Namibia (which includes the uninhabitable Namib Desert, and which has the lowest population density in Africa), is one of the world's leading diamond producers, but it too has major economic problems in the form of high unemployment, rampant inflation, and external debt.

ANGOLA
(ang-GOH-luh)

Population	12,264,000 *(25 per square mile)*
Language(s)	Portuguese, Bantu, *& several other indigenous languages*
Area	481,354 *square miles*
Capital	**Luanda** (loo-AHN-duh)
Economy	$4,500 *annual per capita GDP*
Sovereignty	Republic *independent since 1975*

ZAMBIA
(ZAM-bee-uh)

Population	11,477,000 *(39 per square mile)*
Language(s)	English, Bemba, Kaonda, Lozi, Lunda, Luvale, Nyanja, Tonga, & *many others*
Area	290,586 *square miles*
Capital	**Lusaka** (loo-SAHK-uh)
Economy	$1,000 *annual per capita GDP*
Sovereignty	Republic *independent since 1964*

MALAWI
(muh-LAH-wee)

Population	13,603,000 *(297 per square mile)*
Language(s)	English & Chichewa
Area	45,747 *square miles*
Capital	**Lilongwe** (lee-LAHNG-way)
Economy	$600 *annual per capita GDP*
Sovereignty	Democratic Republic *independent since 1964*

MOZAMBIQUE
(moh-zamm-BEEK)

Population	20,906,000 *(68 per square mile)*
Language(s)	Portuguese *plus several indigenous languages*
Area	309,496 *square miles*
Capital	**Maputo** (mah-POOT-oh)
Economy	$1,500 *annual per capita GDP*
Sovereignty	Republic *independent since 1975*

NAMIBIA
(nah-MIB-ee-uh)

Population	2,055,000 *(6 per square mile)*
Language(s)	English, Afrikaans, & German *plus several indigenous languages*
Area	317,818 *square miles*
Capital	**Windhoek** (VINT-hook)
Economy	$7,500 *annual per capita GDP*
Sovereignty	Republic *independent since* 1990

BOTSWANA
(baht-SWAHN-uh)

Population	1,816,000 *(8 per square mile)*
Language(s)	English & Setswana
Area	224,607 *square miles*
Capital	**Gaborone** (gahb-uh-ROH-nee)
Economy	$10,900 *annual per capita GDP*
Sovereignty	Democratic Republic *independent since* 1966

ZIMBABWE
(zim-BAHB-way)

Population	12,311,000 *(82 per square mile)*
Language(s)	English, Shona, & Sindebele
Area	150,873 *square miles*
Capital	**Harare** (huh-RAH-ray) [*formerly known as* Salisbury]
Economy	$2,100 *annual per capita GDP*
Sovereignty	Democratic Republic *independent since* 1965

SOUTH AFRICA

Population	43,998,000 *(93 per square mile)*
Language(s)	English, Afrikaans, *and 9 other official languages*
Area	470,693 *square miles*
Capital	**Pretoria** (prih-TOR-ree-uh) [Pretoria *is the administrative capital,* Cape Town *is the legislative capital, and* Bloemfontein *is the judicial capital*]
Economy	$13,300 *annual per capita GDP*
Sovereignty	Republic *independent since* 1910

LESOTHO
(luh-SOH-toh *or* luh-SOO-too)

Population	2,125,000 *(181 per square mile)*
Language(s)	Sesotho, English, Zulu, & Xhosa
Area	11,720 *square miles*
Capital	**Maseru** (mahz-uh-roo)
Economy	$2,700 *annual per capita GDP*
Sovereignty	Constitutional Monarchy *independent since 1966*

SWAZILAND

Population	1,133,000 *(169 per square mile)*
Language(s)	English & siSwati
Area	6,704 *square miles*
Capital	**Mbabane** (muh-buh-BAH-nay) [Mbabane *is the administrative capital,* Lobamba *is the royal and legislative capital*]
Economy	$5,300 *annual per capita GDP*
Sovereignty	Monarchy, *independent since 1968*

Indian Ocean Islands

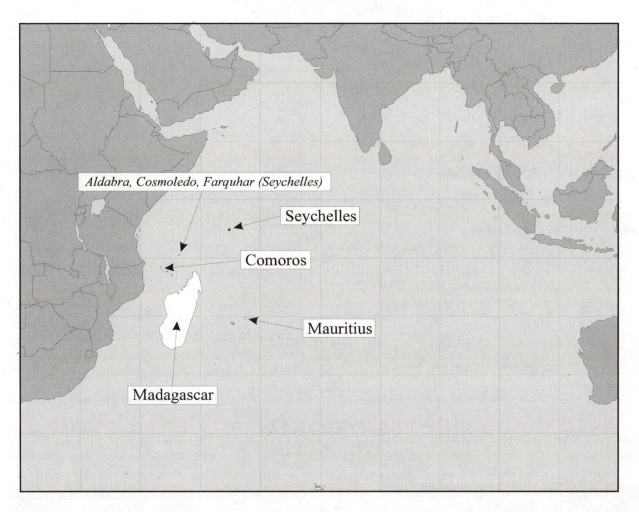

Aldabra, Cosmoledo, Farquhar (Seychelles)

Seychelles

Comoros

Mauritius

Madagascar

There are four independent island countries located in the Indian Ocean: Madagascar, Comoros, Mauritius, and Seychelles. Madagascar is the very large island off the continent's southeast coast. With some eighteen ethnic groups, the country has considerable cultural diversity. Its geographic location has played a large role in the evolution of its historical identity (the Malagasy language, interestingly, is related to the indigenous languages of Indonesia on the other side of the Indian Ocean). As a result of its geographic isolation, Madagascar has many species of plants and animals that are found nowhere else in the world.

The Union of the Comoros is located in the Mozambique Channel west of Madagascar. A former French colony, the Comoros include three main islands (the fourth island in the archipelago, Mayotte, remains under French control, and will be covered in *Appendix A*). Mauritius is located 500 miles east of Madagascar in the southwest Indian Ocean (northeast of the French dependency of Reunion). The island was under French control during most of the 18th century and British control during most of the 19th. Mauritius has the highest population density of any African country. The Seychelles consist of more than 90 islands located about five degrees south of the Equator and approximately 600 miles northeast of Madagascar (the Seychelles also include the Aldabra Islands and the atolls of Cosmoledo and Farquhar off the Tanzanian coast, just north of the tenth parallel). Most of the population is of mixed French and African ancestry; tourism forms an important part of the national economy.

MADAGASCAR

Population	19,449,000 *(86 per square mile)*
Language(s)	Malagasy & French
Area	226,658 *square miles*
Capital	**Antananarivo** (an-tan-nuh-NAHR-ree-voh)
Economy	$900 *annual per capita GDP*
Sovereignty	Republic *independent since* 1960

COMOROS
(KUH-muh-rohz)

Population	711,000 *(824 per square mile)*
Language(s)	Arabic, French, & Comoran (*a mixture of* Arabic *and* Swahili)
Area	863 *square miles*
Capital	**Moroni**
Economy	$600 *annual per capita GDP*
Sovereignty	Republic *independent since* 1975

MAURITIUS
(muh-RISH-us)

Population	1,251,000 *(1,588 per square mile)*
Language(s)	English, French, Creole, Hindi, Urdu, Hakka, & Bojpoori
Area	788 *square miles*
Capital	**Port Louis**
Economy	$13,700 *annual per capita GDP*
Sovereignty	Parliamentary Democracy *independent since* 1968

SEYCHELLES
(say-SHELZ)

Population	82,000 *(466 per square mile)*
Language(s)	English, French, & Creole
Area	176 *square miles*
Capital	**Victoria**
Economy	$7,800 *annual per capita GDP*
Sovereignty	Republic *independent since* 1976

Review Exercises
for
Sub-Saharan Africa

Population Language Area Capital Economy Sovereignty

Country & Capital Identification

1. List the *capital city* of each of the 47 independent countries of Sub-Saharan Africa, and then indicate the location of each country on the map on the opposite page.

Country	Capital	Country	Capital
Angola		Liberia	
Benin		Madagascar	
Botswana		Malawi	
Burkina Faso		Mali	
Burundi		Mauritania	
Cameroon		Mauritius	
Cape Verde		Mozambique	
Cent. African Rep.		Namibia	
Chad		Niger	
Comoros		Nigeria	
Congo		Rwanda	
Congo Republic		São Tomé & Príncipe	
Côte d'Ivoire		Senegal	
Djibouti		Seychelles	
Equatorial Guinea		Sierra Leone	
Eritrea		Somalia	
Ethiopia		South Africa	
Gabon		Swaziland	
Gambia		Tanzania	
Ghana		Togo	
Guinea		Uganda	
Guinea-Bissau		Zambia	
Kenya		Zimbabwe	
Lesotho			

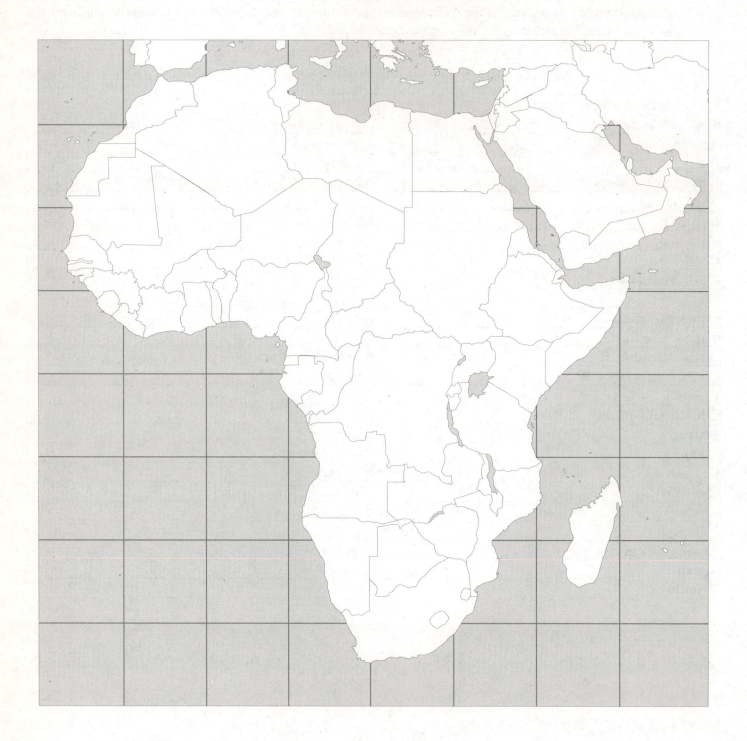

Population

2. Which of the 47 independent countries of Sub-Saharan Africa has the *largest population*?

    _____

3. Which of the 47 independent countries of Sub-Saharan Africa has the *smallest population*?

    _____

4. Which of the 47 independent countries of Sub-Saharan Africa has the *highest population density*?

    _____

5. Which of the 47 independent countries of Sub-Saharan Africa has the *lowest population density*?

    _____

Indicate the location of each of those countries on the map on the opposite page.

Area

6. Which of the 47 independent countries of Sub-Saharan Africa has the *largest area*?

    _____

7. Which of the 47 independent countries of Sub-Saharan Africa has the *smallest area*?

    _____

Indicate the location of each of those countries on the map on the opposite page.

Economy

8. Which of the 47 independent countries of Sub-Saharan Africa has the *highest per capita GDP?*

    _____

9. Which of the 47 independent countries of Sub-Saharan Africa has the *lowest per capita GDP*?

    _____

Indicate the location of each of those countries on the map on the opposite page.

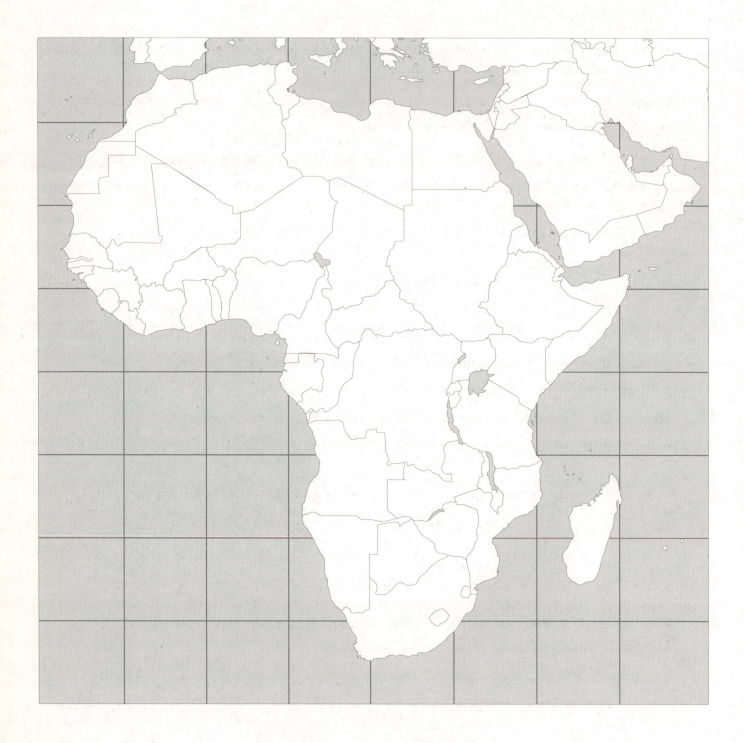

Languages

10.   A very large number of indigenous languages are spoken by the people of Africa, but the legacy of European colonialism is still strikingly apparent in the fact that virtually all African countries include at least one European language as an official or principal language (English, French, and Portuguese are especially prevalent). There is only one exception among the 47 independent countries of Sub-Saharan Africa; name that one African country that is not listed as having a European language as one of its official or principal languages: _____

Indicate the location of that country on the map on the opposite page.

Sovereignty

11.   At the beginning of the 20th century, only 2 of the present 47 independent countries of Africa South of the Sahara were independent. Name those two countries.

_____ &

_____

Indicate the location of each of those countries on the map on the opposite page.

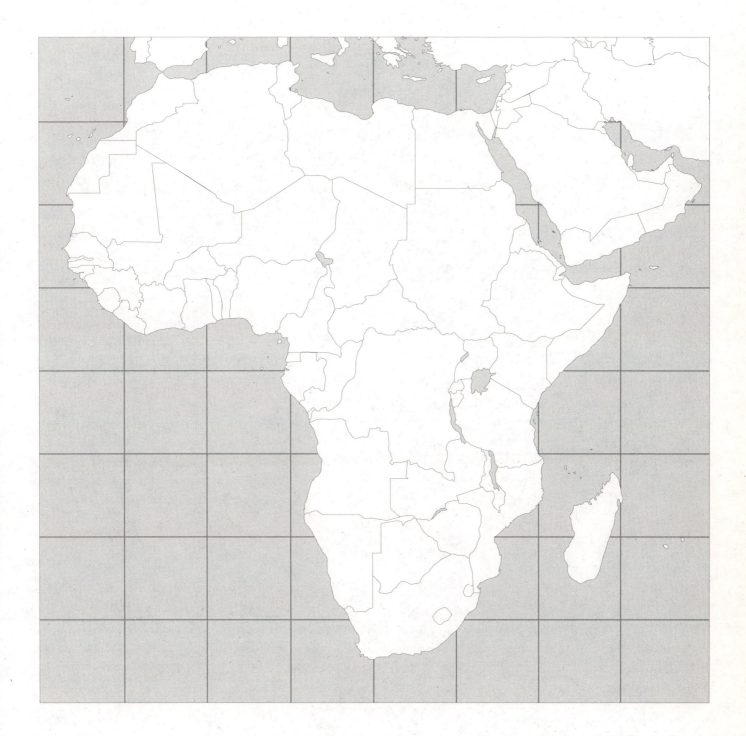

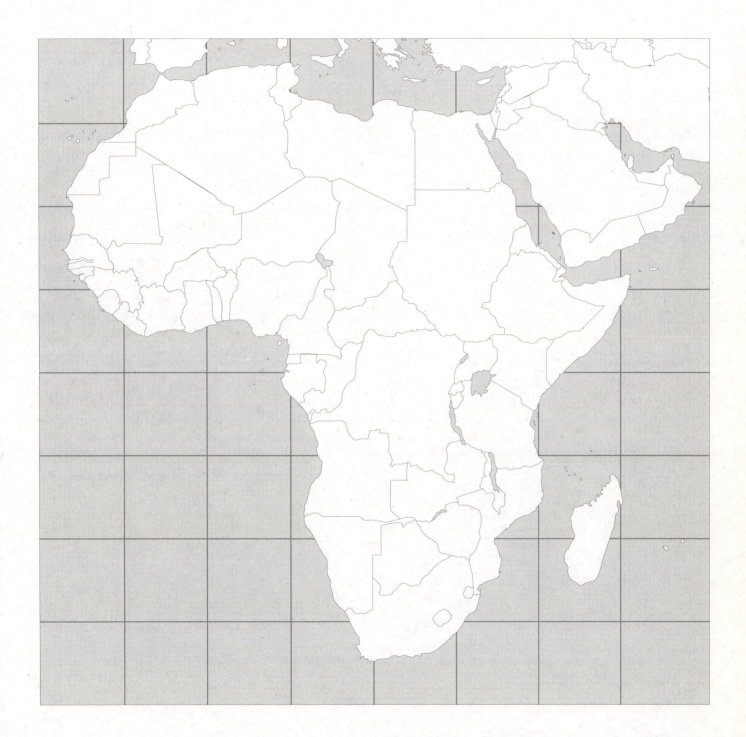

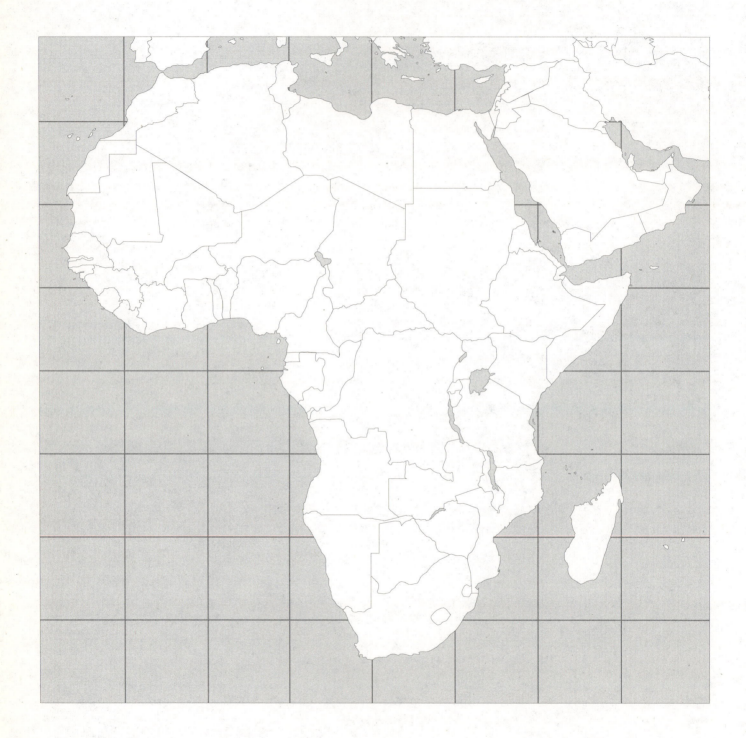

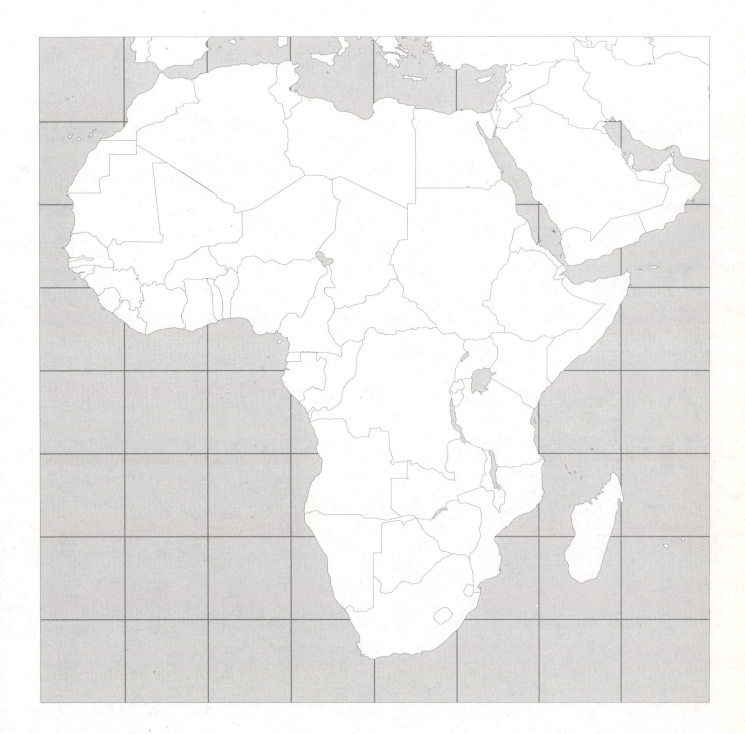

REGION 7
Latin America

Latin America

There are 33 independent countries in Latin America:

Antigua & Barbuda	**Costa Rica**	**Guyana**	**Peru**
Argentina	**Cuba**	**Haiti**	**St. Kitts & Nevis**
Bahamas	**Dominica**	**Honduras**	**St. Lucia**
Barbados	**Dominican Republic**	**Jamaica**	**St. Vincent & Grenadines**
Belize	**Ecuador**	**Mexico**	**Suriname**
Bolivia	**El Salvador**	**Nicaragua**	**Trinidad & Tobago**
Brazil	**Grenada**	**Panama**	**Uruguay**
Chile	**Guatemala**	**Paraguay**	**Venezuela**
Colombia			

The 33 independent countries of Latin America covered in this chapter can be conveniently divided into three broad spatial groups: **Mexico & Central America**, the **Caribbean & Circum-Caribbean**, and **South America**.

Mexico & Central America	Mexico, Guatemala, Belize, El Salvador, Honduras, Nicaragua, Costa Rica, Panama
The Caribbean & Circum-Caribbean	Cuba, Jamaica, Haiti, Dominican Republic, St. Kitts & Nevis, Antigua & Barbuda, Dominica, St. Lucia, Barbados, St. Vincent/Grenadines, Grenada, Trinidad/Tobago, Bahamas
South America	Colombia, Ecuador, Peru, Chile, Argentina, Uruguay, Paraguay, Bolivia, Brazil, Venezuela, Guyana, Suriname

The 13 independent countries of the Caribbean & Circum-Caribbean can be further subdivided into four spatial areas: the **Greater Antilles** (the four largest Caribbean islands, where four sovereign states are found), the **Leeward Islands** (the northern half of the Lesser Antilles, or smaller islands of the Caribbean, where two countries are located), the **Windward Islands** (the southern half of the Lesser Antilles, home to six countries), and the **Circum-Caribbean** (the region immediately north of the Caribbean Sea, where one archipelago nation can be found).

The term **Latin America** refers in the broadest sense to all of the land in the New World portion of the Western Hemisphere that lies south and southeast of the United States. As the term suggests, most of the countries in this region have important cultural and historical ties with the Latin countries of Spain and Portugal. Roman Catholicism, for example, is the dominant religion in the region. Portuguese is spoken in the largest country in Latin America, Brazil, and Spanish is spoken in the majority of the remaining countries. There are pockets in the region that are little touched by Latin culture: English and Dutch are both spoken in some places in the Caribbean and South America, for example. Nevertheless, the term *Latin America* has gained wide currency as an umbrella term for the entire region.

The term **Central America** is reserved for that portion of North America which lies between the southern border of Mexico and the northern border of Colombia. There are seven countries in Central America, ranging from Guatemala in the north to Panama in the south. **Middle America** (also sometimes called **Mesoamerica**) generally refers to Mexico, Central America, and the islands of the Caribbean and Circum-Caribbean. The result of all this nomenclature is a series of embedded terms. *Central America* is part of *Middle America* which in turn is part of *North America*. *Latin America* is a portion of the *New World* that includes all of *South America* plus the southern portion of *North America* (excluding the United States, Canada, and Greenland).

The **Caribbean Sea** lies entirely between the Tropic of Cancer and the Equator. Moving clockwise from the north, this tropical arm of the North Atlantic is bounded by the Greater Antilles, the Lesser Antilles, South America, and Central America. The Caribbean Sea is virtually identical in size to the Mediterranean Sea; the two of them share the top spot as the largest sea in the world. There are 13 independent island nations in the Caribbean and the southern portion of the North Atlantic (*i.e.*, the Circum-Caribbean), ranging from the northernmost Bahamas to the southernmost Trinidad and Tobago.

South America extends from the tropical waters of the Caribbean Sea to the frigid waters of the Drake Passage between Tierra del Fuego and Antarctica. The continent stretches from the Colombian border with Panama in the north to Cape Horn at the tip of Chile in the south. South America is bordered by the Atlantic Ocean to the east and the Pacific Ocean to the west. South America ranks 4th in size among the world's seven continents in terms of area, and 5th in size in terms of population. There are 12 independent countries on the continent of South America (and one political dependency, French Guiana, that will be described in *Appendix A*). The mental image that many people in the United States have of South America's location is misleading in one important respect: the continent is not only *south* of North America, but predominantly *east* as well. The 80° West line of longitude, for example, intersects the *west* coast of South America in Ecuador and Peru and passes through Pittsburgh, Pennsylvania in the *eastern* United States.

The maps on the following two pages depict the locations of the 33 Latin American countries. Given the scale of the maps, however, much of the detail is lost, especially for the islands in the Caribbean. For example, the U.S. and British Virgin Islands are part of an archipelago that includes nearly fifty islands, but the entire archipelago is represented by only two small islands on the map. The multiple-island nations of St. Kitts & Nevis, Antigua & Barbuda, and St. Vincent & the Grenadines are each represented on the map by a single island, and many political dependencies are entirely absent (including Aruba, the Cayman Islands, Anguilla, Montserrat, Saba, St. Barts, St. Eustatius, and St. Martin). Be sure to examine your atlas closely for a complete and accurate picture of the cartography of Latin America.

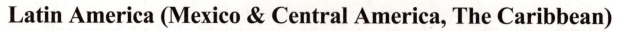

Mexico

Belize

Honduras

Nicaragua

Costa Rica

Guatemala

El Salvador

Panama

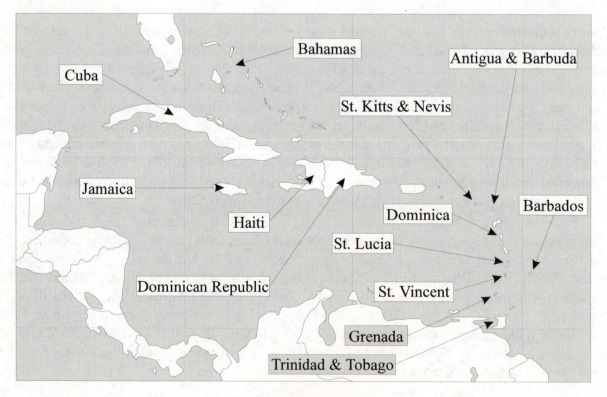

Bahamas

Antigua & Barbuda

Cuba

St. Kitts & Nevis

Jamaica

Dominica

Barbados

Haiti

St. Lucia

Dominican Republic

St. Vincent

Grenada

Trinidad & Tobago

Latin America (South America)

Mexico & Central America

Although much smaller in land area than its two giant neighbors to the north, Mexico is still a very large country. It is larger in both area and population, for example, than any European country other than Russia. In fact, Mexico is more than three times larger than Ukraine, which is the largest country lying entirely within the traditional boundaries of the European continent. At the same time, Mexico is a relatively poor country, with a per capita gross domestic product well below that of most European countries.

Mexico City, which is built upon the foundations of the Aztec capital of Ténochtitlan, is far and away the largest city in the country. In fact, it could arguably be the largest city in the world. During the past several decades, Mexico City has experienced extremely rapid growth from internal migration. The pace of growth has outstripped the city's ability to provide basic services, and large numbers of people live with inadequate supplies of drinking water, sewage facilities, and electricity. Located in a high mountain valley, the Mexican capital is entirely surrounded by towering peaks which act to trap atmospheric pollution and circumscribe urban expansion. As a result, Mexico City has the worst smog of any city in the Americas and some of the worst traffic congestion of any city in the world.

The seven nations of Central America are all small countries; each, in fact, is smaller in both area and population than the state of Florida. There is a wide range in population densities, however, from sparsely-inhabited Belize, with around 30 people per square mile, to overcrowded El Salvador, with more than 800 people per square mile.

Most of the countries of Central America, along with Mexico, are Spanish-speaking. In terms of ethnicity, most, again like Mexico, are predominantly Mestizo, exhibiting a discernible mixture of American Indian and European ancestry. The principal exceptions are English-speaking Belize, with a population that is primarily of African descent, and Costa Rica, with a population that is primarily of European descent.

Other than the former British colony of Belize (which was previously known as British Honduras and which gained its independence in 1981), the countries of Central America were all once part of the Spanish colonial empire. With the exception of Panama (which gained its independence from Colombia in 1903), all of the remaining countries on the isthmus became independent in the first half of the 19th century. Costa Rica, however, is the only former Spanish colony in Central America to have enjoyed a long history of peaceful democratic government and to have escaped the devastating effects of civil war and military dictatorship. Largely as a result, Costa Rica has the highest standard of living and the largest middle class of any Central American country.

262   Places of the World

MEXICO

Population	108,701,000 *(143 per square mile)*
Language(s)	Spanish *plus* Mayan, Nahuatl, *and various other indigenous languages*
Area	758,452 *square miles*
Capital	**Mexico City**
Economy	$10,700 *annual per capita GDP*
Sovereignty	Federal Democratic Republic *independent since* 1810

GUATEMALA
(gwut-uh-MAH-lah)

Population	12,728,000 *(303 per square mile)*
Language(s)	Spanish *and more than 20 Amerindian languages*
Area	42,042 *square miles*
Capital	**Guatemala City**
Economy	$5,000 *annual per capita GDP*
Sovereignty	Democratic Republic *independent since* 1839

BELIZE
(buh-LEEZ)

Population	294,000 *(33 per square mile)*
Language(s)	English *plus* Spanish, Mayan, Garifuna, & Creole
Area	8,867 *square miles*
Capital	**Belmopan** (bell-moh-PAN)
Economy	$8,400 *annual per capita GDP*
Sovereignty	Parliamentary Democracy *independent since* 1981

EL SALVADOR

Population	6,948,000 *(855 per square mile)*
Language(s)	Spanish
Area	8,124 *square miles*
Capital	**San Salvador**
Economy	$4,900 *annual per capita GDP*
Sovereignty	Republic *independent since* 1841

HONDURAS

Population	7,484,000 *(173 per square mile)*
Language(s)	Spanish
Area	43,277 *square miles*
Capital	**Tegucigalpa** (teh-goo-see-GAHL-puh)
Economy	$3,100 *annual per capita GDP*
Sovereignty	Democratic Republic *independent since* 1839

NICARAGUA
(nick-uh-RAH-gwuh)

Population	5,675,000 *(113 per square mile)*
Language(s)	Spanish
Area	50,054 *square miles*
Capital	**Managua** (mah-NAH-gwuh)
Economy	$3,100 *annual per capita GDP*
Sovereignty	Republic *independent since* 1838

COSTA RICA

Population	4,134,000 *(210 per square mile)*
Language(s)	Spanish
Area	19,730 *square miles*
Capital	**San José**
Economy	$12,500 *annual per capita GDP*
Sovereignty	Democratic Republic *independent since* 1821

PANAMA

Population	3,242,000 *(111 per square mile)*
Language(s)	Spanish
Area	29,157 *square miles*
Capital	**Panama City**
Economy	$8,200 *annual per capita GDP*
Sovereignty	Democratic Republic *independent since* 1903

The Caribbean & Circum-Caribbean: The Greater Antilles

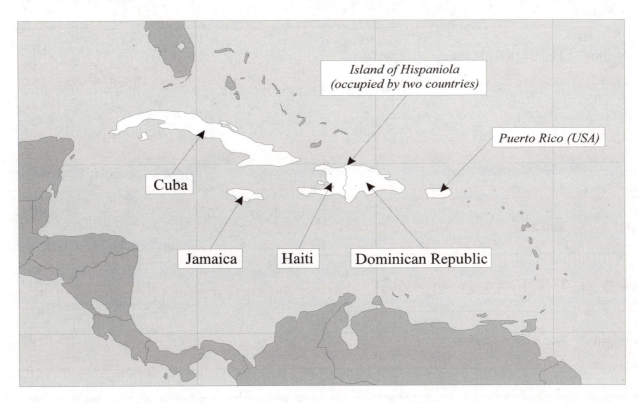

The Greater Antilles are comprised of the four largest islands in the Caribbean: Cuba, Jamaica, Hispaniola, and Puerto Rico. Cuba is the largest sovereign state in the Caribbean, both in terms of area and population, although Hispaniola, which is shared by the independent countries of Haiti and Dominican Republic, has the largest total population of any single island. (Puerto Rico, which is a Commonwealth of the United States and not an independent country, will be covered in *Appendix A*.)

Like the majority of the islands in the Caribbean, the Greater Antilles are of volcanic origin, and have mountainous and hilly terrains as a result. There are a handful of low-lying coral islands in the Caribbean, such as the Cayman Islands and Anegada, and a relatively small number of islands that are part of the continental shelf of South America, such as Aruba, Bonaire, and Curaçao, but all of the Greater Antilles and most of the Lesser Antilles were created by volcanic activity that began about 25 million years ago. That activity continues in a few places today; the island of Montserrat was largely depopulated in 1997 following a series of violent eruptions.

The Caribbean region as a whole is sometimes referred to as the West Indies, but the remarkable degree of cultural, linguistic, and political diversity within the area makes it impossible to speak of a single "West Indian" identity. The four independent countries of the Greater Antilles illustrate the Caribbean's wide variety. Both Cuba and Dominican Republic are Spanish-speaking societies comprised primarily of people of mixed European and African ancestry, but the former has an authoritarian socialist government while the latter has enjoyed several decades of democratic rule. English-speaking Jamaica, whose population is predominantly of African descent, is a parliamentary democracy, while French-and-Creole-speaking Haiti, one of the poorest countries in the Western Hemisphere, is struggling to emerge from decades of brutal dictatorial regimes.

CUBA

Population	11,394,000 *(266 per square mile)*
Language(s)	Spanish
Area	42,804 *square miles*
Capital	**Havana**
Economy	$4,100 *annual per capita GDP*
Sovereignty	Socialist Republic *independent since* 1902

JAMAICA

Population	2,780,000 *(655 per square mile)*
Language(s)	English
Area	4,244 *square miles*
Capital	**Kingston**
Economy	$4,700 *annual per capita GDP*
Sovereignty	Parliamentary Democracy *independent since* 1962

HAITI

Population	8,706,000 *(813 per square mile)*
Language(s)	French & Creole
Area	10,714 *square miles*
Capital	**Port-au-Prince** (pohrt-oh-PRANTS)
Economy	$1,800 *annual per capita GDP*
Sovereignty	Republic *independent since* 1804

DOMINICAN REPUBLIC

Population	9,366,000 *(500 per square mile)*
Language(s)	Spanish
Area	18,730 *square miles*
Capital	**Santo Domingo**
Economy	$8,400 *annual per capita GDP*
Sovereignty	Democratic Republic *independent since* 1844

The Lesser Antilles: The Leeward Islands & the Windward Islands

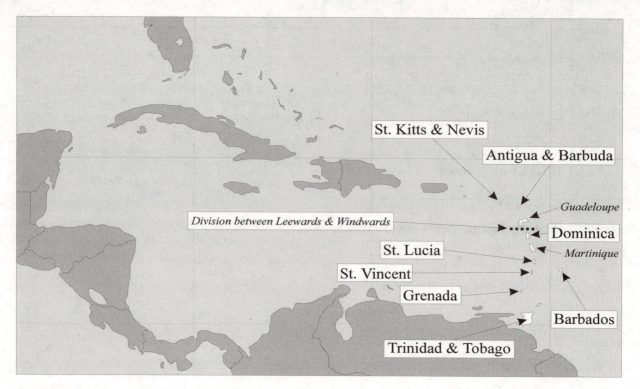

The eastern boundary of the Caribbean Sea is formed by a chain of generally small islands, collectively referred to as the Lesser Antilles, which extends from the island of St. Thomas just east of Puerto Rico to the island of Trinidad just north of Venezuela. The Lesser Antilles are divided into two groups: the northern portion is called the Leeward Islands (pronounced "LOO-uhrd") while the southern potion is known as the Windward Islands. The boundary between the Leeward and Windward Islands falls between the islands of Guadeloupe and Dominica.

There are only eight sovereign states in the Lesser Antilles, and all eight are English-speaking members of the Commonwealth (see *Appendix B*). The remainder of the islands in the eastern Caribbean are dependencies of the United States, the United Kingdom, France, and the Netherlands (all of those dependencies will be reviewed in *Appendix A*). Two of the eight independent countries in the Lesser Antilles, St. Kitts & Nevis and Antigua & Barbuda, fall within the Leeward Islands; both rank among the ten smallest countries in the world in terms of population, and both are extremely small in land area as well. All of the sovereign states of the Leeward and Windward Islands gained their independence in the 1960's, 1970's, or 1980's.

Six of the eight independent countries of the Lesser Antilles fall within the Windward Islands: Dominica, St. Lucia, Barbados, St. Vincent & the Grenadines, Grenada, and Trinidad & Tobago. The populations of each of the sovereign states in the Windward Islands are predominantly of African ancestry, with the single exception of Trinidad & Tobago, whose population is comprised of nearly equal mixtures of African and East Indian lineages. All of the Windward Islands, like the majority of islands throughout the Caribbean, have relatively underdeveloped economies, although the tourism industry is continuing to grow throughout the region. With nearly 1,700 people per square mile, Barbados has the highest population density of any country in North or South America.

ST. KITTS AND NEVIS
(NEE-viss)

Population	39,000 *(386 per square mile)*
Language(s)	English
Area	101 *square miles*
Capital	**Basseterre** (bahs-tair) [rhymes with "hair"]
Economy	$8,200 *annual per capita GDP*
Sovereignty	Parliamentary Democracy *independent since 1983*

ANTIGUA AND BARBUDA
(ahn-TEEG-uh *and* bahr-BOOD-uh)

Population	69,000 *(404 per square mile)*
Language(s)	English
Area	171 *square miles*
Capital	**St. John's**
Economy	$10,900 *annual per capita GDP*
Sovereignty	Parliamentary Democracy *independent since 1981*

DOMINICA
(dahm-ih-NEE-kuh)

Population	72,000 *(248 per square mile)*
Language(s)	English
Area	290 *square miles*
Capital	**Roseau**
Economy	$3,800 *annual per capita GDP*
Sovereignty	Democratic Republic *independent since* 1978

ST. LUCIA
(LOO-shuh)

Population	171,000 *(718 per square mile)*
Language(s)	English
Area	238 *square miles*
Capital	**Castries**
Economy	$4,800 *annual per capita GDP*
Sovereignty	Parliamentary Democracy *independent since* 1979

BARBADOS
(bar-BAY-dohss)

Population	281,000 *(1,693 per square mile)*
Language(s)	English
Area	166 *square miles*
Capital	**Bridgetown**
Economy	$18,400 *annual per capita GDP*
Sovereignty	Parliamentary Democracy *independent since* 1966

ST. VINCENT AND THE GRENADINES
(gren-uh-DEENZ)

Population	118,000 *(787 per square mile)*
Language(s)	English
Area	150 *square miles*
Capital	**Kingstown**
Economy	$3,600 *annual per capita GDP*
Sovereignty	Parliamentary Democracy *independent since* 1979

GRENADA
(gruh-NAY-duh)

Population	90,000 *(677 per square mile)*
Language(s)	English
Area	133 *square miles*
Capital	**St. George's**
Economy	$3,900 *annual per capita GDP*
Sovereignty	Parliamentary Democracy *independent since 1974*

TRINIDAD AND TOBAGO
(toh-BAY-goh)

Population	1,057,000 *(534 per square mile)*
Language(s)	English *plus* Hindi, French, Spanish, & Chinese
Area	1,980 *square miles*
Capital	**Port-of-Spain**
Economy	$19,800 *annual per capita GDP*
Sovereignty	Parliamentary Democracy *independent since 1962*

Places of the World

Circum-Caribbean

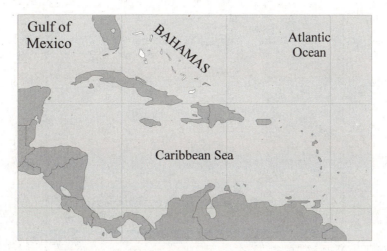

      Although many American tourists associate the Bahamas with the Caribbean, the entire country actually lies outside the Caribbean Sea. Indeed, the Bahamas are in many ways very distinct from their island neighbors to the south. While the majority of the islands in the Caribbean are volcanic and mountainous, the low-lying Bahamas are of coral origin. Unlike the islands of the Caribbean, which fall entirely within the tropics, the Bahamas straddle the Tropic of Cancer; the capital of Nassau is subtropical, and the island of Grand Bahama has even seen brief snow flurries in extreme winters. Thriving banking and tourism industries give the Bahamas a higher per capita gross domestic product than any of the independent countries of Mesoamerica.

BAHAMAS

Population	306,000 *(57 per square mile)*	
Language(s)	English	
Area	5,382 *square miles*	
Capital	**Nassau** (NASS-aw)	
Economy	$21,600 *annual per capita GDP*	
Sovereignty	Parliamentary Democracy *independent since* 1973	

South America

South America is characterized by striking geographic diversity. The topography includes equatorial rainforests, barren deserts, rolling plains, and rugged mountains, while the climate ranges from the steamy heat of the tropics to the frigid gales of the sub-Antarctic. The towering Andes mountains form the western spine of the continent and include ten peaks that are higher than any of those found in North America. The eastern half of the continent is dominated by the immense Amazon River basin, the largest in the world.

The cultural diversity of the region is equally impressive. The country of Brazil (which accounts for half of South America's population and area) is notable for its relatively harmonious blending of European, Amerindian, African, and Asian races and cultures. The Andean countries of Bolivia, Peru, and Ecuador include significant numbers of indigenous Indians who have maintained their distinctive languages and traditions. At the same time, there are important commonalties throughout the region. Nine of the twelve independent countries in South America share the same language, Spanish, and Roman Catholicism is the dominant religion on the continent.

The continent as a whole has a relatively small population and a relatively low population density (Ecuador's population density approximates the worldwide average, but the other eleven South American countries are all well below average in the number of people per square mile). The bulk of the continent's population lives in the coastal areas; much of the interior is sparsely inhabited. Most of the countries in the region are experiencing high rates of internal migration, as increasing numbers of people move from rural to urban areas.

Most of the independent countries of South America achieved their sovereignty in the first half of the 19th century (the two exceptions, Guyana and Suriname, both became independent in the second half of the 20th century). There is one political dependency remaining on the continental mainland, French Guiana, and one offshore possession, the Falkland Islands (claimed by both the United Kingdom and Argentina, although under British control); both of those entities will be covered in *Appendix A*.

COLOMBIA

Population	44,380,000 *(101 per square mile)*
Language(s)	Spanish
Area	439,737 *square miles*
Capital	**Bogota** (boh-goh-TAH) [*officially* Santa Fe de Bogotá]
Economy	$8,600 *annual per capita GDP*
Sovereignty	Republic *independent since* 1819

ECUADOR
(ECK-wah-dor)

Population	13,756,000 *(126 per square mile)*
Language(s)	Spanish
Area	109,484 *square miles*
Capital	**Quito** (KEE-toh)
Economy	$4,500 *annual per capita GDP*
Sovereignty	Republic *independent since* 1830

PERU

Population	28,675,000 (*58 per square mile*)
Language(s)	Spanish & Quechua
Area	496,225 *square miles*
Capital	**Lima** (LEE-muh)
Economy	$6,600 *annual per capita GDP*
Sovereignty	Republic *independent since* 1821

CHILE

Population	16,286,000 (*56 per square mile*)
Language(s)	Spanish
Area	291,930 *square miles*
Capital	**Santiago**
Economy	$12,600 *annual per capita GDP*
Sovereignty	Republic *independent since* 1818

ARGENTINA

Population	40,302,000 *(38 per square mile)*
Language(s)	Spanish
Area	1,073,519 *square miles*
Capital	**Buenos Aires**
Economy	$15,200 *annual per capita GDP*
Sovereignty	Republic *independent since* 1816

URUGUAY

Population	3,461,000 *(51 per square mile)*
Language(s)	Spanish
Area	67,574 *square miles*
Capital	**Montevideo** (moan-tay-vee-DAY-oh)
Economy	$10,900 *annual per capita GDP*
Sovereignty	Republic *independent since* 1825

PARAGUAY

Population	6,669,000 *(42 per square mile)*
Language(s)	Spanish
Area	157,048 *square miles*
Capital	**Asunción** (ah-soohn-see-OHN)
Economy	$4,800 *annual per capita GDP*
Sovereignty	Republic *independent since* 1811

BOLIVIA

Population	9,119,000 *(21 per square mile)*
Language(s)	Spanish, Aymara, and Quechua
Area	424,165 *square miles*
Capital	**Sucre** (SOO-kray, the capital) and **La Paz** (lah PAHZ, the seat of government)
Economy	$3,100 *annual per capita GDP*
Sovereignty	Republic *independent since* 1825

BRAZIL

Population	190,011,000 *(58 per square mile)*
Language(s)	Portuguese
Area	3,300,172 *square miles*
Capital	**Brasilia**
Economy	$8,800 *annual per capita GDP*
Sovereignty	Federal Republic *independent since* 1822

VENEZUELA

Population	26,024,000 *(74 per square mile)*
Language(s)	Spanish
Area	352,145 *square miles*
Capital	**Caracas** (kah-RAH-kuss)
Economy	$7,200 *annual per capita GDP*
Sovereignty	Federal Republic *independent since* 1811

GUYANA
(guy-AHNA)

Population	769,000 *(9 per square mile)*
Language(s)	English *plus* Creole, Hindi, Urdu, *and various Amerindian languages*
Area	83,000 *square miles*
Capital	**Georgetown**
Economy	$4,900 *annual per capita GDP*
Sovereignty	Republic *independent since* 1966

SURINAME
(suhr-uh-NAHM)

Population	471,000 *(7 per square mile)*
Language(s)	Dutch *plus* English, Sranang Tongo, Hindustani, & Javanese
Area	63,037 *square miles*
Capital	**Paramaribo** (pah-rah-MAH-ree-boh)
Economy	$7,100 *annual per capita GDP*
Sovereignty	Democratic Republic *independent since* 1975

Review Exercises
for
Latin America

Population Language Area Capital Economy Sovereignty

Country & Capital Identification

1. List the *capital city* of each of the 33 independent countries of Latin America, and then indicate the location of each country on the maps on the following two pages.

Country	Capital
Antigua & Barbuda	
Argentina	
Bahamas	
Barbados	
Belize	
Bolivia	
Brazil	
Chile	
Colombia	
Costa Rica	
Cuba	
Dominica	
Dominican Republic	
Ecuador	
El Salvador	
Grenada	
Guatemala	
Guyana	
Haiti	
Honduras	
Jamaica	
Mexico	
Nicaragua	
Panama	
Paraguay	
Peru	
St. Kitts & Nevis	
St. Lucia	
St. Vincent & Grenadines	
Suriname	
Trinidad & Tobago	
Uruguay	
Venezuela	

2. The *division between the Leeward and Windward Islands* of the Lesser Antilles occurs immediately north of the island of _____ and immediately south of the island of _____ (draw a line at that point on the appropriate map).

Population

3.   Which of the 33 independent countries of Latin America has the *largest population*?

4.   Which of the 33 independent countries of Latin America has the *smallest population*?

5.   Which of the 33 independent countries of Latin America has the *highest population density*?

6.   Which of the 33 independent countries of Latin America has the *lowest population density*?

Indicate the location of each of those countries on the maps on the following two pages

Area

7.   Which of the 33 independent countries of Latin America has the *largest area*?

8.   Which of the 33 independent countries of Latin America has the *smallest area*?

Indicate the location of each of those countries on the maps on the following two pages.

Economy

9.   Which of the 33 independent countries of Latin America has the *highest per capita GDP*?

10.   Which of the 33 independent countries of Latin America has the *lowest per capita GDP*?

Indicate the location of each of those countries on the maps on the following two pages.

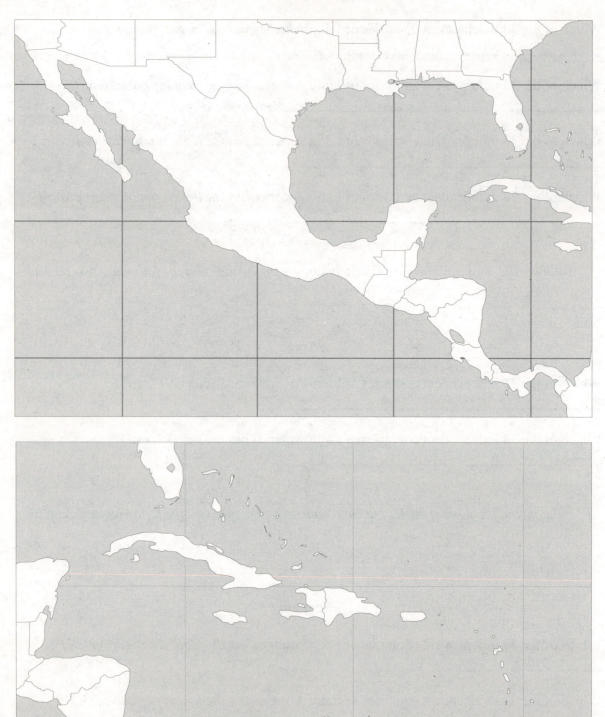

Languages

11.   The two independent Caribbean countries that are *officially Spanish-speaking* are:
_____ & _____.

12.   The only independent Caribbean country that has something *other than English or Spanish* as its official or principal language is _____.

13.   The only three South American countries that *do not have Spanish* as one of their official languages are:
_____, _____,
& _____.

Indicate the location of each of those countries on the maps on the following two pages.

Sovereignty

14.   The *only socialist republic* among the 33 sovereign states of Latin America is:
_____.

Indicate the location of that country on the appropriate map on the following two pages.

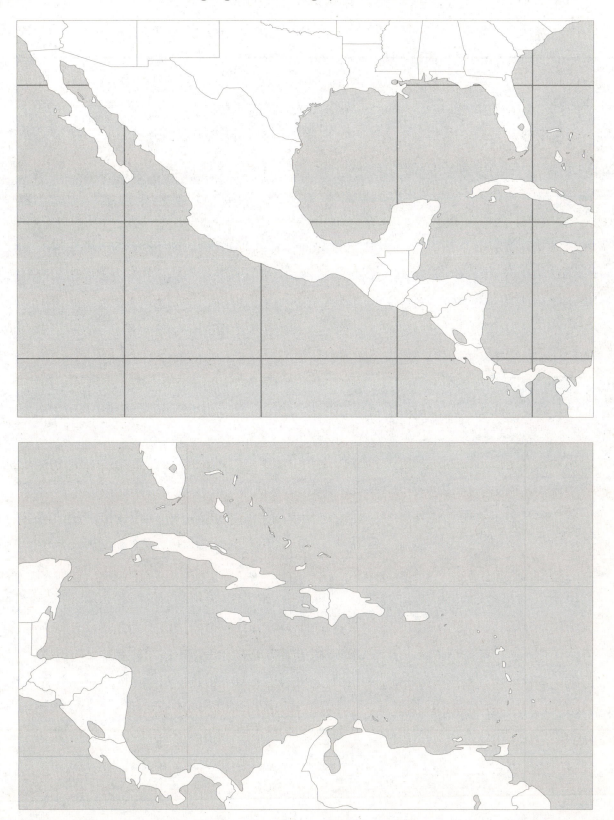

Places of the World

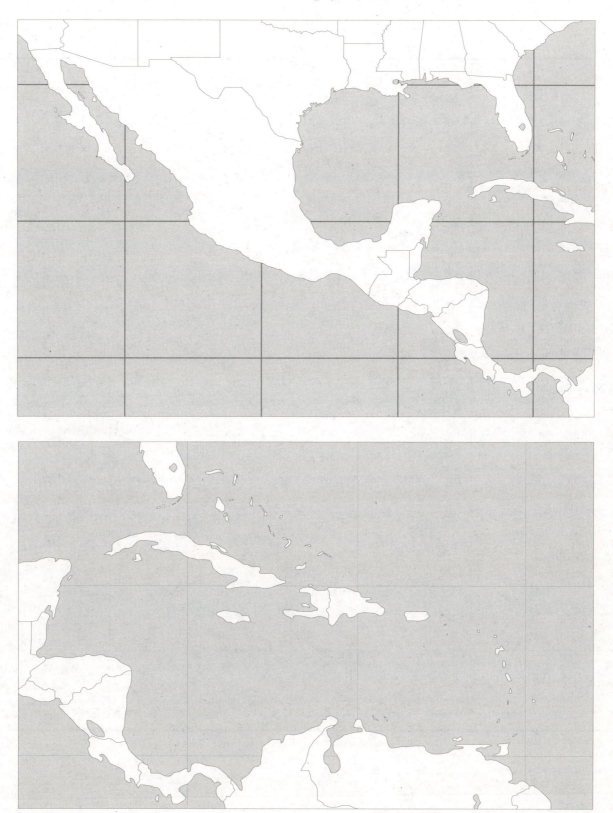

Places of the World

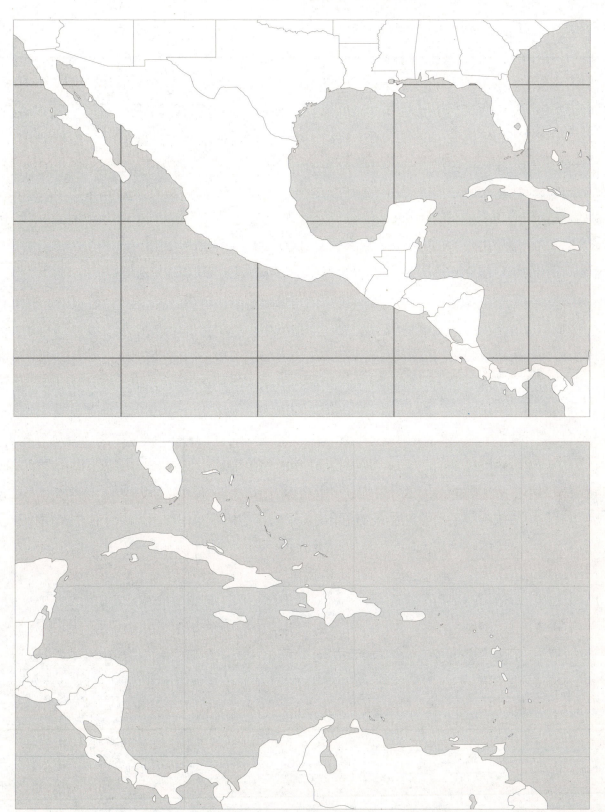

Places of the World

REGION 8:
United States & Canada

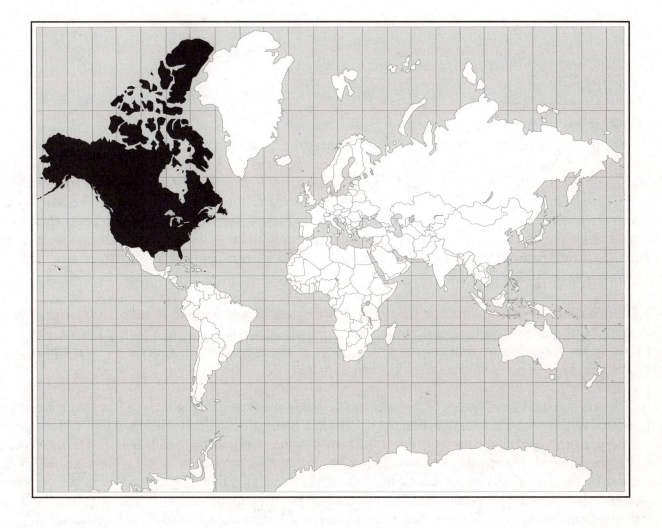

United States and Canada

There are 2 independent countries in North America that are covered in this chapter:

United States of   **Canada**
America

The traditional continental borders of **North America** extend from north of the Arctic Circle to south of the Tropic of Cancer. The continent stretches from Greenland and the islands of the Canadian Arctic to the boundary with South America at the border between Panama and Colombia. North America is bordered by the Atlantic Ocean to the east, the Pacific Ocean to the west, and the Arctic Ocean to the north. The islands of the Caribbean are generally included within the traditional boundaries of North America, as are the Atlantic islands of Bermuda and the Bahamas (Greenland and Bermuda are discussed in *Appendix A*; the Bahamas were covered in *Region 7*). North America ranks 3rd in size among the world's seven continents in terms of area, and 4th in size in terms of population.

This chapter focuses on that portion of North America which is sometimes called Anglo-America and which consists of the two countries of Canada and the United States. The Anglo-Saxon influence was highly significant in the formation of both countries, but growing ethnic diversity in Canada and the United States makes the term *Anglo-America* increasingly less appropriate.

Canada and the United States of America are both enormous in geographic area (they rank second and third, respectively, among the world's countries). Canada's population is relatively small, given the country's vast territory, so its population density is one of the lowest in the world. Both Canada and the United States are affluent countries, with per capita gross domestic products that rank among the world's highest. The USA, of course, has by far the world's largest economy. In terms of the total value of goods and services exchanged, Canada and the United States have the largest trading partnership of any two countries on earth. Not surprisingly, Canada and the U.S. share the longest undefended border of any two countries in the world.

In this brief chapter, you will be asked to identify the internal political divisions within both the United States and Canada. Much of this material is likely to be familiar to you already; at the very least, you probably devoted some time in the earliest stages of your academic career to learning the 50 states and their capital cities. Helpful hints for organizing the U.S. states and Canadian provinces are described below.

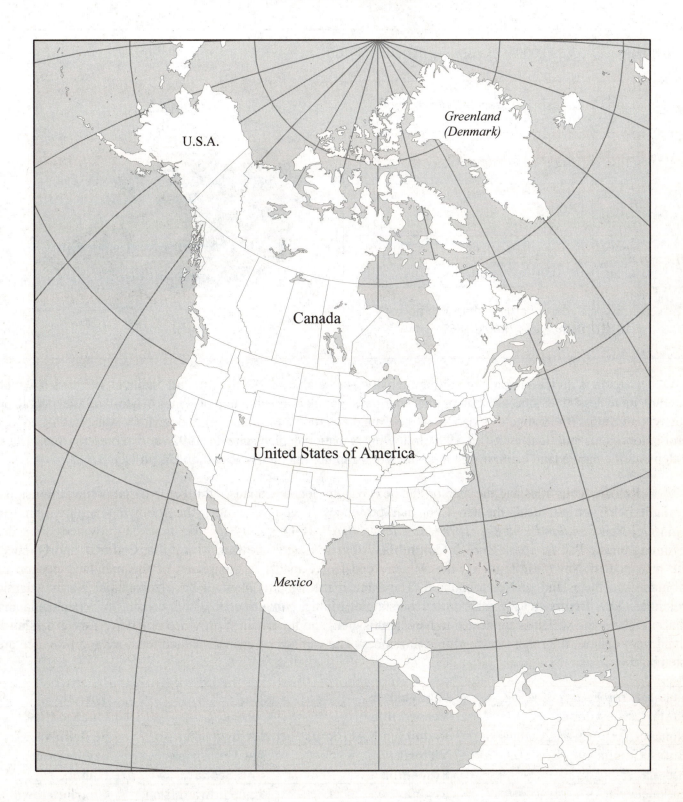

U.S.A.

Greenland
(Denmark)

Canada

United States of America

Mexico

CANADA

Population	33,390,000 (*9 per square mile*)
Language(s)	English and French *(both official)*
Area	3,855,103 *square miles*
Capital	**Ottawa**
Economy	$35,700 *annual per capita GDP*
Sovereignty	Federal Parliamentary Democracy *independent since* 1867

Canada is divided into three northern territories (the Yukon, Northwest, and Nunavut Territories, which together have less than one percent of the country's population) and ten provinces. Most of the people in Nunavut are Inuits (the indigenous people of the Canadian arctic and subarctic); the territory was established as a semi-autonomous political entity in 1999 following a national plebiscite in 1992, and represents the largest settlement of a native land claim in history. Nunavut's capital, Iqaluit, is located on Baffin Island.

Refer to your atlas and the map on the next page to locate Canada's thirteen internal political divisions. It might help you to recall the ten Canadian provinces if you remember the following acronym-phrase: *BASMOQ-Newfoundland and the Three Maritimes*. **BASMOQ** stands for the first six provinces in order beginning on the Pacific coast: **British Columbia**, **Alberta**, **Saskatchewan**, **Manitoba**, **Ontario**, and **Quebec**. The province of **Newfoundland**, on the Atlantic coast of Canada, is comprised of the mainland portion of Labrador and the island of Newfoundland. The three maritime provinces all have substantial North Atlantic coastlines: **New Brunswick**, which borders the state of Maine, **Nova Scotia**, which consists of a peninsula and adjacent island and which is connected to New Brunswick, and the smallest province of **Prince Edward Island**, which lies offshore the other two maritime provinces. The capitals of each Canadian territory and province are listed below:

Yukon Territory	**Whitehorse**	Ontario	**Toronto**
Northwest Territories	**Yellowknife**	Quebec	**Quebec City**
Nunavut Territory (NOO-nuh-voot)	**Iqaluit** (ee-KAL-oo-it)	Newfoundland	**St. John's**
British Columbia	**Victoria**	New Brunswick	**Fredericton**
Alberta	**Edmonton**	Nova Scotia	**Halifax**
Saskatchewan	**Regina**	Prince Edward Island	**Charlottetown**
Manitoba	**Winnipeg**		

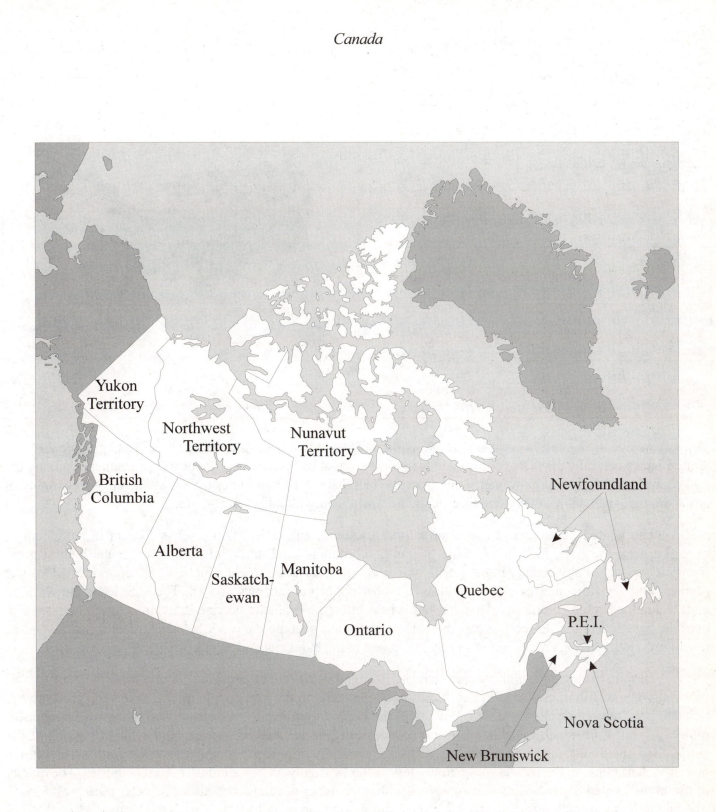

UNITED STATES OF AMERICA

Population	301,140,000 *(79 per square mile)*
Language(s)	English
Area	3,794,083 *square miles*
Capital	**Washington, District of Columbia**
Economy	$43,800 *annual per capita GDP*
Sovereignty	Federal Democratic Republic *independent since* 1776

As one of the largest countries in the world, both in terms of geographic area and demographic size, the United States naturally plays a major role on the world's stage. Its political and economic preeminence among the 194 countries of the world, however, stems from the fact that the United States possesses the largest economy of any sovereign state while at the same time being the world's only military superpower.

The United States enjoys abundant natural resources, although it is not self-sufficient in petroleum or several critical minerals, including bauxite, platinum, chromium, and tin. While it has by far the highest gross domestic product of any country in the world, it does not quite have the world's highest *per capita* gross domestic product, and several countries in the world have higher per capita incomes as well. The U.S. leads the world in high technology research and development, but it has suffered a dramatic decline in manufacturing employment over the past generation. In recent years the United States has become the world's leading debtor nation, and its rate of individual savings has fallen to among the lowest in the industrialized world.

Despite having an extremely high standard of living by most measures, the country is not without notable problems. The distribution of wealth has become more unequal in the past couple of decades as the gap between rich and poor has appreciably widened. Poverty in the U.S. is significantly correlated with ethnicity: members of minority groups are much more likely to be economically disadvantaged. Among the affluent nations of the world, the United States is unique in having a large, chronically impoverished underclass. Largely as a result, the rate of violent crime in the U.S. is substantially higher than virtually all other industrialized countries. The quality of life in the United States suffers in comparison with many other developed countries by other measures as well. The rate of infant mortality is relatively high, for example, and the rate of literacy is relatively low. While the U.S. can still boast the world's finest university system, the country's public education system is seriously deficient in comparison with most other affluent countries.

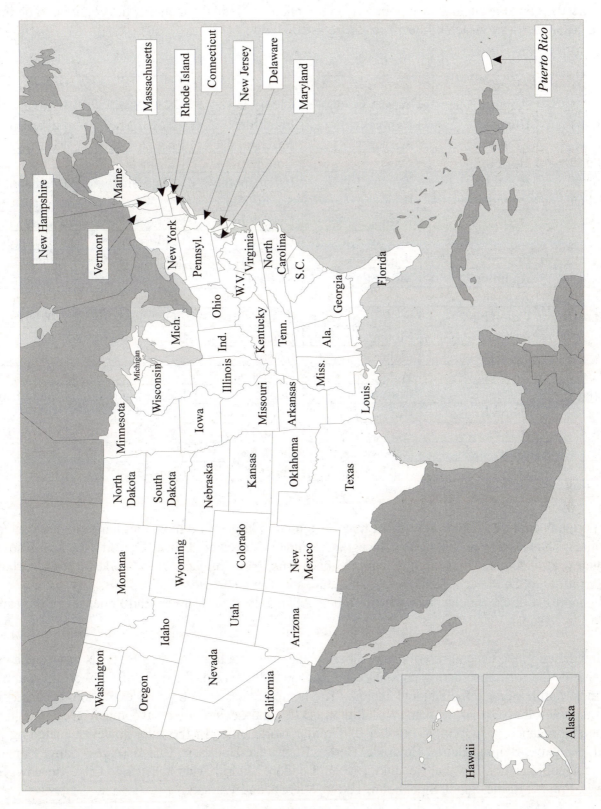

Massachusetts
Rhode Island
Connecticut
New Jersey
Delaware
Maryland
Puerto Rico
New Hampshire
Maine
Vermont
New York
Pennsyl.
W.V. Virginia
North Carolina
S.C.
Ohio
Kentucky
Tenn.
Georgia
Florida
Mich.
Ind.
Ala.
Michigan
Wisconsin
Illinois
Missouri
Arkansas
Miss.
Louis.
Minnesota
Iowa
North Dakota
South Dakota
Nebraska
Kansas
Oklahoma
Texas
Montana
Wyoming
Colorado
New Mexico
Washington
Idaho
Utah
Arizona
Oregon
Nevada
California
Hawaii
Alaska

The 50 states are generally placed into seven recognizable geographic regions, based largely upon historical criteria. These regional groupings also point to cultural and economic distinctions that continue to be important today. The capital cities for each of the 50 states are listed below:

New England		Midwest		Mid-Atlantic	
Maine	Augusta	West Virginia	Charleston	New York	Albany
New Hampshire	Concord	Kentucky	Frankfort	New Jersey	Trenton
Vermont	Montpelier	Missouri	Jefferson City	Pennsylvania	Harrisburg
Massachusetts	Boston	Ohio	Columbus	Delaware	Dover
Connecticut	Hartford	Indiana	Indianapolis	Maryland	Annapolis
Rhode Island	Providence	Illinois	Springfield		
		Iowa	Des Moines	Great Plains	
		Michigan	Lansing	North Dakota	Bismarck
South		Wisconsin	Madison	South Dakota	Pierre
Virginia	Richmond	Minnesota	St. Paul	Nebraska	Lincoln
North Carolina	Raleigh			Kansas	Topeka
South Carolina	Columbia	West		Oklahoma	Oklahoma City
Georgia	Atlanta	Montana	Helena		
Florida	Tallahassee	Idaho	Boise		
Alabama	Montgomery	Wyoming	Cheyenne	Pacific	
Mississippi	Jackson	Colorado	Denver	California	Sacramento
Louisiana	Baton Rouge	Utah	Salt Lake City	Oregon	Salem
Texas	Austin	Nevada	Carson City	Washington	Olympia
Arkansas	Little Rock	New Mexico	Santa Fe	Alaska	Juneau
Tennessee	Nashville	Arizona	Phoenix	Hawaii	Honolulu

You may find it helpful to remember that the country's largest and most familiar cities are generally *not* state capitals. Thus Baltimore, Charlotte, Chicago, Cincinnati, Cleveland, Dallas, Detroit, Houston, Kansas City, Los Angeles, Philadelphia, Pittsburgh, Miami, Minneapolis, New Orleans, New York, Norfolk, Orlando, St. Louis, San Antonio, San Diego, San Francisco, Seattle, and Tampa are *not* state capitals. It is easier to recall the relatively short list of exceptions among major cities: Atlanta, Boston, Denver, and Phoenix *are* the capitals of their respective states.

Incidentally, the general rule for remembering state capitals in the United States is exactly the opposite of the general rule for remembering national capitals around the world: for most of the world's countries, the largest, most familiar city *is* the national capital. (The United States is an obvious exception, because there are many U.S. cities that are larger than Washington, D.C., and several that have much greater cultural and economic significance; a substantial number of familiar world cities fit the rule, however, including Athens, Baghdad, Bangkok, Beijing, Beirut, Belgrade, Berlin, Bucharest, Budapest, Buenos Aires, Cairo, Copenhagen, Damascus, Dublin, Hanoi, Havana, Helsinki, Lisbon, London, Madrid, Manila, Mexico City, Moscow, Nairobi, Oslo, Paris, Prague, Rome, Seoul, Stockholm, Tokyo, Vienna, and Warsaw.)

Places of the World

Review Exercises
for
United States & Canada

Population Language Area Capital Economy Sovereignty

1. List the *capital cities* of each of the 50 states of the United States, and then indicate the location of each state on the map on the opposite page.

State	Capital	State	Capital
Alabama		Montana	
Alaska		Nebraska	
Arizona		Nevada	
Arkansas		New Hampshire	
California		New Jersey	
Colorado		New Mexico	
Connecticut		New York	
Delaware		North Carolina	
Florida		North Dakota	
Georgia		Ohio	
Hawaii		Oklahoma	
Idaho		Oregon	
Illinois		Pennsylvania	
Indiana		Rhode Island	
Iowa		South Carolina	
Kansas		South Dakota	
Kentucky		Tennessee	
Louisiana		Texas	
Maine		Utah	
Maryland		Vermont	
Massachusetts		Virginia	
Michigan		Washington	
Minnesota		West Virginia	
Mississippi		Wisconsin	
Missouri		Wyoming	

2. What is the national *capital* of the United States of America?

   _____

3. What is the *approximate population* of the United States?

   _____

4. What is the *approximate population density* of the United States?

   _____

5. What is the *approximate area* of the United States?

   _____

6. What is the *approximate per capita GDP* of the United States?

   _____

7. What is the *type of sovereignty* of the United States?

   _____

308   Places of the World

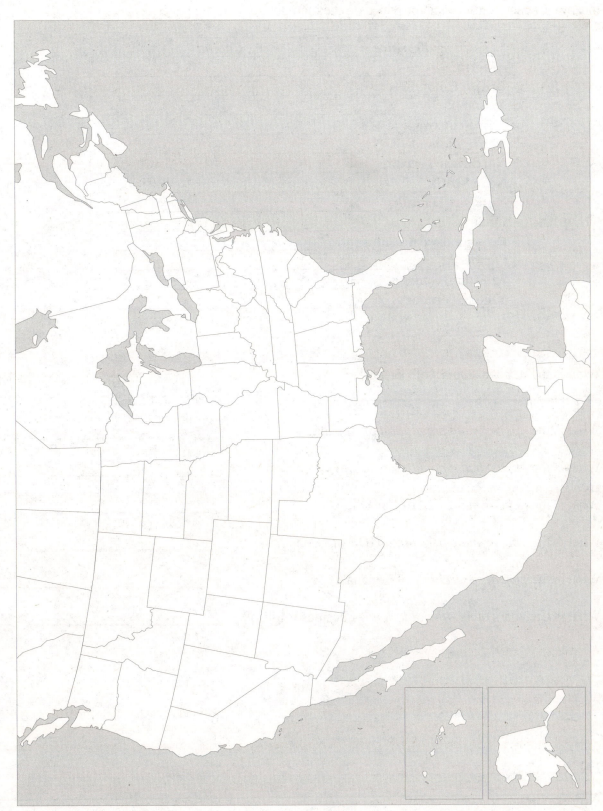

8.　List the *capital cities* of each of the 3 territories and 10 provinces of Canada, and then indicate the location of each territory and province on the map on the opposite page.

Territory/Province	Capital
Alberta	
British Colombia	
Manitoba	
New Brunswick	
Newfoundland	
Northwest Territory	
Nova Scotia	
Nunavut Territory	
Ontario	
Prince Edward Island	
Quebec	
Saskatchewan	
Yukon Territory	

9.　What is the national *capital* of Canada?

10.　What is the *approximate population* of Canada?

11.　What is the *approximate population density* of Canada?

12.　What is the *approximate area* of Canada?

13.　What is the *approximate per capita GDP* of Canada?

14.　What is the *type of sovereignty* of Canada?

15.　What is Canada's other *official language* besides English?

Places of the World

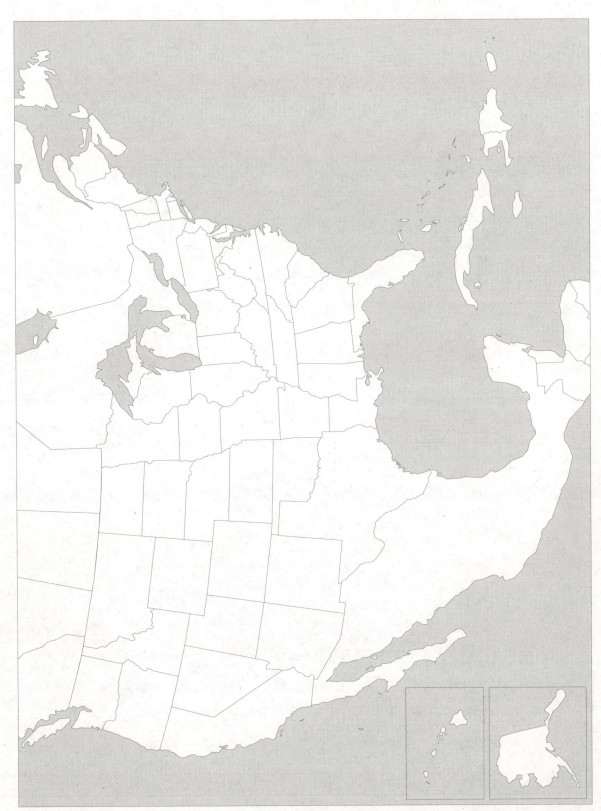

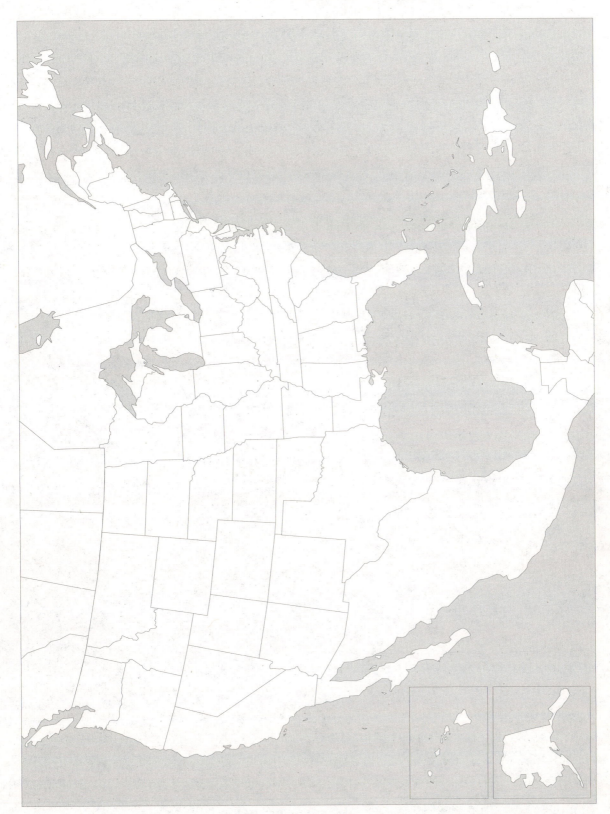

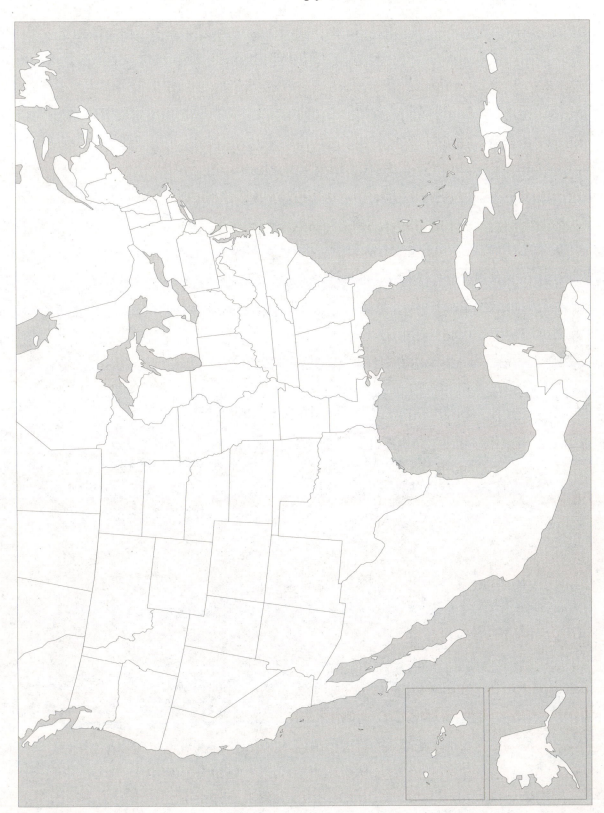

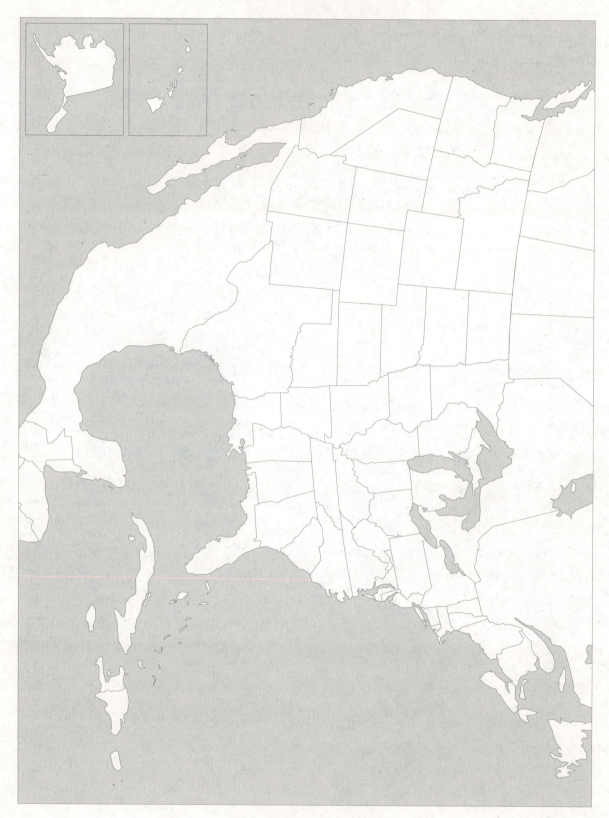

APPENDIX A
Political Dependencies and Antarctica

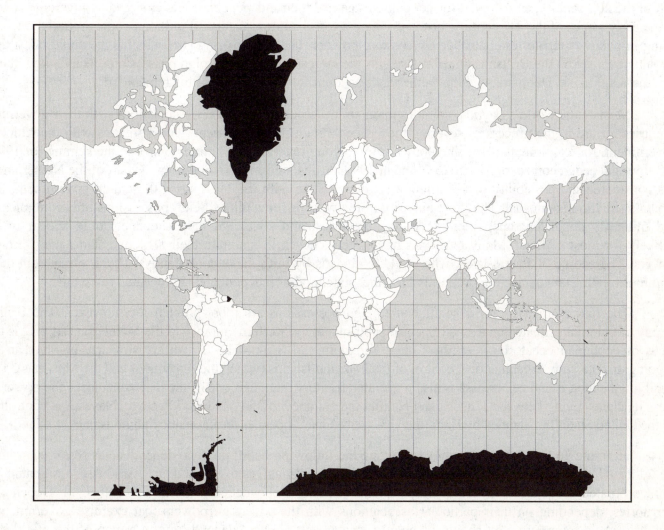

Political Dependencies

The era of European colonialism that began in the early 16th century came to a virtual end in the second half of the 20th century, as scores of former colonies gained their independence and emerged as sovereign states. In the early stages of the 21st century, however, there are still approximately 10 million people around the world living in overseas territories controlled by Western powers. The United Kingdom, which once could boast that the sun never set on the British Empire, still retains the largest number of foreign possessions. The United States of America, France, Denmark, and the Netherlands account for the majority of the remainder.

Most of the world's *de facto* colonies maintain peaceful and voluntary associations with their governing countries, or are in the process of severing those ties through mutual consent. Under treaties with the Chinese government, for example, the British colony of Hong Kong reverted to Chinese control in the summer of 1997, and Portugal ceded control of Macau to China in 1999. Aruba, currently an autonomous part of the Netherlands, has considered the possibility of becoming a sovereign state with the blessing of the Dutch government, but Aruba's situation is unusual. The great majority of the world's remaining foreign possessions are economically and militarily dependent upon their governing countries, and sovereign independence would appear to be unlikely for most of them. Many territories have freely chosen to maintain their affiliated status with foreign powers. The people of Puerto Rico, who are U.S. citizens by birth, have consistently elected to maintain their Commonwealth status, rejecting proposals for either statehood or independence in repeated referenda.

The future prospects for a handful of foreign possessions are more problematic, however. The British colony of Gibraltar, situated on a small Spanish peninsula at the western entrance to the Mediterranean Sea, is the focus of continued debate between the United Kingdom and Spain. The British are similarly involved in an ongoing dispute with Argentina over control of the Falkland Islands (which the Argentines call the Malvinas). The small uninhabited island of Navassa is also a disputed territory, although one of the disputants lacks the power to press its claim. Lying between Jamaica and Haiti at the entrance to the Windward Passage, Navassa is controlled by the United States but claimed by Haiti (the U.S. Coast Guard operates an automatic lighthouse on the island).

Examine your atlas closely while reviewing the following list of political dependencies scattered around the globe. These foreign possessions have a range of political statuses, and are variously designated as Unincorporated Territories, Overseas Departments, Crown Dependencies, Commonwealths, or Dependent Territories, depending on their particular associations with the sovereign powers that exercise governmental authority. As a minimum, you should be able to locate each political dependency on a map of the world and name its sovereign affiliation.

	DEPENDENCY	AFFILIATION	LOCATION
North American Region	Bermuda	United Kingdom	E of North Carolina
	Greenland	Denmark	NE of Canada
	St. Pierre & Miquelon	France	S of Newfoundland Island
	Turks and Caicos	United Kingdom	SE of Bahamas

	DEPENDENCY	AFFILIATION	LOCATION
South American / Caribbean Region	Anguilla	United Kingdom	NNW of St. Kitts & Nevis
	Aruba	Netherlands	N of Venezuela
	Bonaire	Netherlands	N of Venezuela
	British Virgin Islands	United Kingdom	E of Puerto Rico
	Cayman Islands	United Kingdom	S of Cuba
	Curaçao	Netherlands	N of Venezuela
	Falkland Islands	United Kingdom	E of Argentina
	French Guiana	France	E of Suriname
	Galapagos	Ecuador	W of Ecuador
	Guadeloupe	France	N of Dominica
	Martinique	France	S of Dominica
	Montserrat	United Kingdom	SW of Antigua
	Puerto Rico	United States	E of Dominican Republic
	Saba	Netherlands	NW of St. Kitts & Nevis
	St. Barthelemy (a.k.a. St. Barts)	France	NW of St. Kitts & Nevis
	St. Martin	France	NW of St. Kitts & Nevis
	Sint Eustatius	Netherlands	NW of St. Kitts & Nevis
	Sint Maarten	Netherlands	NW of St. Kitts & Nevis
	South Georgia	United Kingdom	SE of Argentina
	U.S. Virgin Islands	United States	E of Puerto Rico

	DEPENDENCY	AFFILIATION	LOCATION
European Region	Azores	Portugal	W of Portugal
	Channel Islands	United Kingdom	English Channel
	Faeroe Islands	Denmark	N of Scotland
	Gibraltar	United Kingdom	SW Mediterranean coast of Spain
	Isle of Man	United Kingdom	Between Great Britain & Ireland

	FORMER DEPENDENCY	AFFILIATION	LOCATION
Asian Region	Hong Kong	Integral Part of China	Southeast coast of China
	Macau	Integral Part of China	Southeast coast of China

	DEPENDENCY	AFFILIATION	LOCATION
African Region	Amsterdam Island	France	ESE of South Africa
	Canary Islands	Spain	W of Morocco
	Kerguelen	France	SE of South Africa
	Madeira	Portugal	W of Morocco
	Mayotte	France	Comoros archipelago
	Reunion	France	SW of Mauritius
	St. Helena	United Kingdom	WNW of Namibia
	Tristan da Cunha	United Kingdom	SW of South Africa

	DEPENDENCY	AFFILIATION	LOCATION
Pacific Region	American Samoa	United States	E of Western Samoa
	Easter Island	Chile	W of Chile
	Guam	United States	E of Philippines
	Midway Islands	United States	NW of Hawaii
	New Caledonia	France	SW of Vanuatu
	Northern Marianas	United States	ENE of Philippines
	Pitcairn Island	United Kingdom	SE of Tonga
	Tahiti	France	ENE of Tonga
	Wake Island	United States	N of Marshall Islands

The map on the opposite page illustrates the approximate location of each of the political dependencies listed above; the locations correspond to the numbers on the map, as indicated in the table below (notice that four of the numbers on the map—5, 11, 12, & 30—indicate the approximate location of more than one political dependency).

Key to Map Locations of Political Dependencies

1	Tahiti	13	St. Pierre & Miquelon	25	South Georgia
2	Pitcairn Island	14	Greenland	26	Reunion
3	Easter Island	15	Faeroe Islands	27	Kerguelen
4	Galapagos Islands	16	Isle of Man	28	Amsterdam Island
5	Aruba, Bonaire, Curaçao	17	Channel Islands	29	Mayotte
6	French Guiana	18	Azores	30	Hong Kong, Macau
7	Puerto Rico	19	Madeira	31	Northern Marianas
8	Cayman Islands	20	Canary Islands	32	Midway Islands
9	Turks & Caicos	21	Gibraltar	33	Wake Island
10	Bermuda	22	St. Helena	34	Guam
11	British Virgins, U.S. Virgins	23	Tristan da Cunha	35	American Samoa
12	Anguilla, Guadeloupe, Martinique, Montserrat, Saba, St. Barts, St. Martin, Sint Eustatius, Sint Maarten	24	Falkland Islands	36	New Caledonia

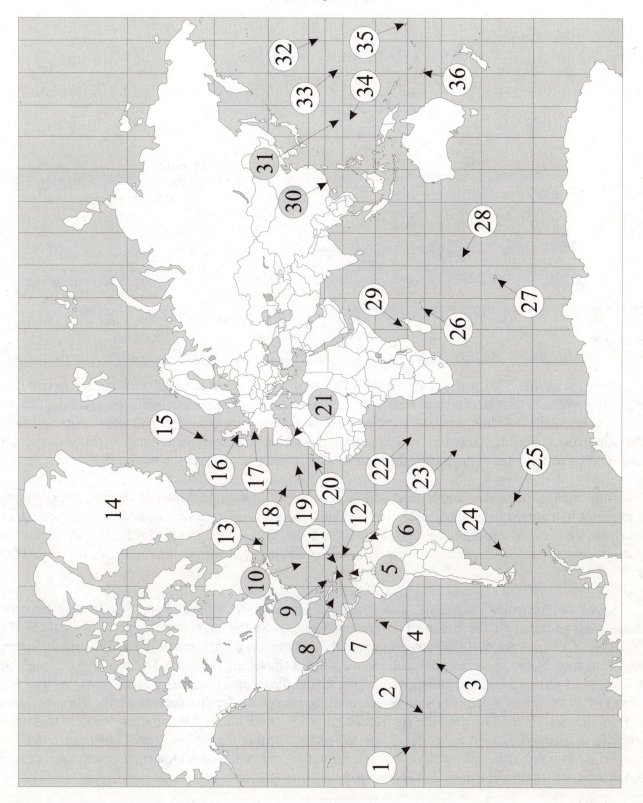

Antarctica

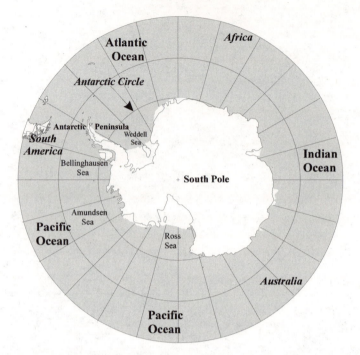

Antarctica ranks 5th in size among the world's seven continents in terms of area, and 7th in size in terms of population. With a total land area of nearly 5.5 million square miles, Antarctica is nearly twice the size of Australia. Other than a handful of research stations scattered around the continent and a few tourist destinations on the Antarctic Peninsula, however, Antarctica has no permanent settlements and virtually no permanent population. The continent of Antarctica is the largest uninhabited land mass in the world.

Antarctica is the coldest and highest continent on earth. In the mid-winter of August, temperatures can drop to minus 95 degrees Fahrenheit, and a 10 foot thick collar of sea ice forms around the continent, doubling its solid surface area. Antarctica's average elevation is 7,000 feet, although up to 4/5th of that consists of the permanent ice cap that covers approximately 95% of the continent; at its maximum, the ice cap is 13,000 feet thick.

Seven of the world's sovereign states claim territory in Antarctica. Four of them are countries in relatively close proximity to the continent (Argentina, Chile, Australia, and New Zealand), while three of them are European countries that were involved in the early exploration of Antarctica in the 19th and 20th centuries (United Kingdom, Norway, and France). All seven countries signed the Antarctic Treaty in 1959, along with the United States of America and several other countries. Under the terms of the treaty, which was renewed in 1991 by a total of 39 countries, Antarctica is to be reserved for peaceful scientific purposes only. For a period of 50 years from the treaty's renewal date, the signatories have agreed to a ban on mineral exploitation in Antarctica, and they have pledged to keep the continent demilitarized. Neither the United States nor the United Nations recognizes the validity of any of the territorial claims on the continent, and all the signatories to the treaty have agreed that no new territorial claims will be asserted.

Review Exercises
for
Political Dependencies & Antarctica

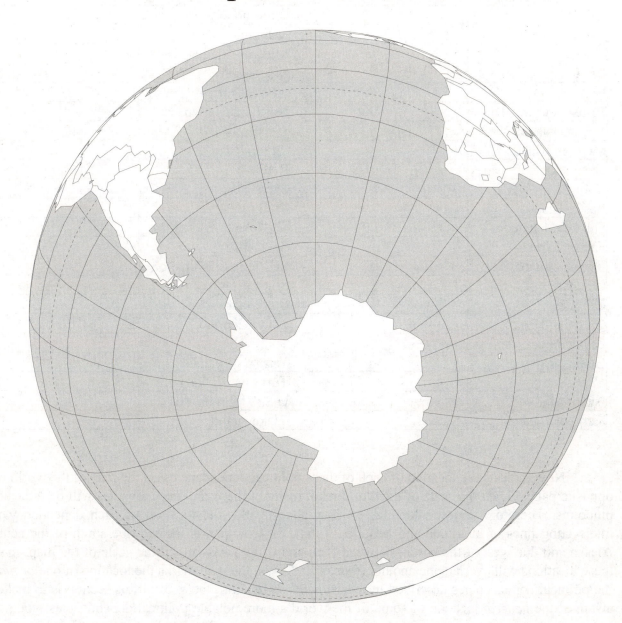

Location & Affiliation of Dependencies, Profile of Antarctica

1.   List the sovereign affiliation for each of the following political dependencies.

Dependency	Sovereign Affiliation	Dependency	Sovereign Affiliation
American Samoa		*Macau*	
Amsterdam Island		Madeira	
Anguilla		Martinique	
Aruba		Mayotte	
Azores		Midway Islands	
Bermuda		Montserrat	
Bonaire		New Caledonia	
British Virgin Islands		Northern Marianas	
Canary Islands		Pitcairn Island	
Cayman Islands		Puerto Rico	
Channel Islands		Reunion	
Curaçao		Saba	
Easter Island		St. Barthelemy	
Falkland Islands		St. Helena	
Faeroe Islands		St. Martin	
French Guiana		St. Pierre & Miquelon	
Galapagos		Sint Eustatius	
Gibraltar		Sint Maarten	
Greenland		South Georgia	
Guadeloupe		Tahiti	
Guam		Tristan da Cunha	
Hong Kong		Turks and Caicos	
Isle of Man		U.S. Virgin Islands	
Kerguelen Island		Wake Island	

Now indicate the location of each of these political dependencies on the map of the world on the opposite page. Given the scale of the world map, many of these tiny dependencies will be indicated by pinpoints. Therefore you should refer to your atlas to plot the location of each dependency using intersecting lines of longitude and latitude. (Tahiti, for example, is located due south of the center of Alaska and due west of the border between Peru and Chile.) Also given the scale of the map, many of these locations will overlap one another (thus each "X" on the map marks the location of one *or more* of the political dependencies described in this *Appendix*). In short, the goal of this section is to make you aware of the *general* location of some of the world's more notable political dependencies; you are not being asked to make exact locational distinctions between immediately adjacent dependencies.

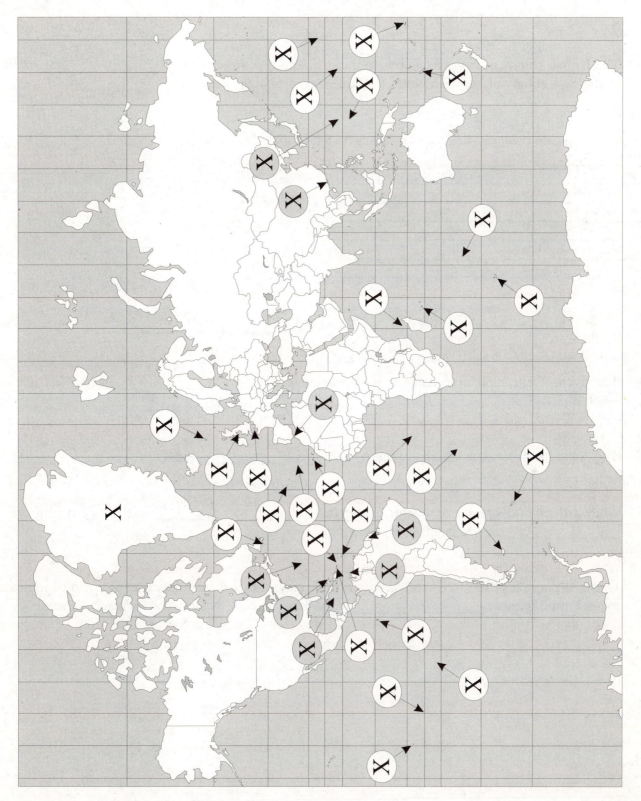

Places of the World

2. The Western power that continues to have the *largest number of political dependencies* in the world is

    _____.

3. The British colony of *Hong Kong* reverted to Chinese control in the year _____, while the
    Portuguese colony of *Macau* reverted to Chinese control in the year _____.

4. List the *seven countries* that claim territory on the continent of Antarctica:

    _____,

    _____,

    _____,

    _____,

    _____,

    _____.

5. In total land area, Antarctica measures approximately _____ miles.

6. The average elevation of Antarctica is approximately _____ feet above sea
    level.

7. The Antarctic Treaty was originally signed in the year _____ and renewed in _____.

8. Among the world's seven continents, Antarctica ranks _____ in size in terms of *area* and
    _____ in size in terms of *population*.

9. Under the terms of the most recent Antarctic Treaty, the ban on mineral exploitation in Antarctica
    extends until the year _____.

APPENDIX B
International Organizations

Dozens of international organizations have been created to address a wide range of political, economic, social, environmental, and regional interests and concerns. The following brief profiles describe a few of the more notable associations of the world's countries.

United Nations (UN)

The United Nations was established in 1945; its aim is to maintain international peace and security and to promote cooperation involving economic, social, cultural, and humanitarian problems. As of this writing in February 2008, there are 192 members of the United Nations—all of the 194 independent countries listed in *Appendix F* except Taiwan and Vatican City. Taiwan is regarded by the UN as part of China, and Vatican City maintains permanent observer status at the UN. The UN Security Council is the organ of the United Nations specifically charged with maintaining international peace and security. There are five permanent members of the Security Council (China, France, Russia, United Kingdom, and United States), each of which has veto power, and ten members elected for two-year terms by the UN General Assembly.

The Commonwealth (*also known as* The Commonwealth of Nations)

The Commonwealth was established in 1931 as a voluntary association of countries formerly associated with the British Empire; its aim is to foster multinational cooperation and assistance among its member states. As of 2008, there are 53 members of The Commonwealth: Antigua & Barbuda, Australia, Bahamas, Bangladesh, Barbados, Belize, Botswana, Brunei, Cameroon, Canada, Cyprus, Dominica, Fiji, Gambia, Ghana, Grenada, Guyana, India, Jamaica, Kenya, Kiribati, Lesotho, Malawi, Malaysia, Maldives, Malta, Mauritius, Mozambique, Namibia, Nauru, New Zealand, Nigeria, Pakistan (suspended), Papua New Guinea, St. Kitts & Nevis, St. Lucia, St. Vincent & the Grenadines, Samoa, Seychelles, Sierra Leone, Singapore, Solomon Islands, South Africa, Sri Lanka, Swaziland, Tanzania, Tonga, Trinidad & Tobago, Tuvalu, Uganda, United Kingdom, Vanuatu, and Zambia. The 2 billion citizens of the Commonwealth represent about 30% of the world's population, and reflect a very wide range of cultures, traditions, religions, and ethnicities.

Commonwealth of Independent States (CIS)

The Commonwealth of Independent States was established in 1991; its aim is to coordinate relations among the member states and to provide a mechanism for the orderly dissolution of the Union of Soviet Socialist Republics. As of 2008, there are 12 members of the CIS: Armenia, Azerbaijan, Belarus, Georgia, Kazakhstan, Kyrgyzstan, Moldova, Russia, Tajikistan, Turkmenistan, Ukraine, and Uzbekistan.

European Union (EU)

The European Union was established in 1993 as the successor to the European Community; its aim is to coordinate policy among its members in the areas of economics, defense, and justice. As of early 2008, there are 27 members of the EU (Austria, Belgium, Bulgaria, Cyprus, Czech Republic, Denmark, Estonia, Finland, France, Germany, Greece, Hungary, Ireland, Italy, Latvia, Lithuania, Luxembourg, Malta, Netherlands, Poland, Portugal, Romania, Slovakia, Slovenia, Spain, Sweden, United Kingdom) and 3 candidates for membership (Croatia, Macedonia, and Turkey). The EU has its own flag, anthem, and currency, and it is increasingly taking on the attributes of a sovereign country.

League of Arab States (LAS—also known as the Arab League)

The League of Arab States was established in 1945; its aim is to promote economic, social, political, and military cooperation among the member states. As of 2008, there are 21 members in the LAS, plus the Palestine Liberation Organization: Algeria, Bahrain, Comoros, Djibouti, Egypt, Iraq, Jordan, Kuwait, Lebanon, Libya, Mauritania, Morocco, Oman, Qatar, Saudi Arabia, Somalia, Sudan, Syria, Tunisia, United Arab Emirates, and Yemen.

North Atlantic Treaty Organization (NATO)

The North Atlantic Treaty Organization was established in 1949; its aim is to promote mutual defense and cooperation among its member states. As of early 2008, NATO includes 26 countries: Belgium, Bulgaria, Canada, Czech Republic, Denmark, Estonia, France, Germany, Greece, Hungary, Iceland, Italy, Latvia, Lithuania, Luxembourg, Netherlands, Norway, Poland, Portugal, Romania, Slovakia, Slovenia, Spain, Turkey, United Kingdom, and United States of America.

Organization of American States (OAS)

The Organization of American States was chartered in 1948 as the successor to the International Union of American Republics; its aim is to promote regional peace and security as well as economic and social development. As of 2008, there are 35 members in the OAS: Antigua & Barbuda, Argentina, Bahamas, Barbados, Belize, Bolivia, Brazil, Canada, Chile, Colombia, Costa Rica, Cuba (excluded from formal participation since 1962), Dominica, Dominican Republic, Ecuador, El Salvador, Grenada, Guatemala, Guyana, Haiti, Honduras, Jamaica, Mexico, Nicaragua, Panama, Paraguay, Peru, St. Kitts & Nevis, St. Lucia, St. Vincent & the Grenadines, Suriname, Trinidad & Tobago, United States of America, Uruguay, and Venezuela.

Organization of Petroleum Exporting Countries (OPEC)

The Organization of Petroleum Exporting Countries was established in 1960; its aim is to coordinate petroleum policies among its members states. As of 2008, there are 11 members in OPEC: Algeria, Indonesia, Iran, Iraq, Kuwait, Libya, Nigeria, Qatar, Saudi Arabia, United Arab Emirates, and Venezuela.

APPENDIX C
Country Rank by Area

Rank	Country	Area (miles^2)	Rank	Country	Area (miles^2)
1	Russia	6,592,849	38	Zambia	290,586
2	Canada	3,855,103	39	Morocco	275,117
3	United States	3,794,083	40	Burma	261,228
4	China	3,690,045	41	Afghanistan	251,773
5	Brazil	3,300,172	42	Somalia	246,201
6	Australia	2,969,910	43	Central African Repub.	240,536
7	India	1,222,510	44	Ukraine	233,090
8	Argentina	1,073,519	45	Madagascar	226,658
9	Kazakhstan	1,049,156	46	Kenya	224,961
10	Sudan	967,500	47	Botswana	224,607
11	Algeria	919,595	48	France	208,482
12	Congo-Kinshasa	905,446	49	Yemen	203,850
13	Saudi Arabia	830,000	50	Thailand	198,115
14	Mexico	758,452	51	Spain	194,885
15	Indonesia	735,310	52	Turkmenistan	188,457
16	Libya	679,362	53	Cameroon	183,568
17	Iran	636,372	54	Papua New Guinea	178,704
18	Mongolia	604,829	55	Sweden	173,732
19	Peru	496,225	56	Uzbekistan	172,742
20	Chad	495,755	57	Iraq	169,235
21	Niger	489,192	58	Paraguay	157,048
22	Angola	481,354	59	Zimbabwe	150,873
23	Mali	478,841	60	Japan	145,850
24	South Africa	470,693	61	Germany	137,847
25	Colombia	439,737	62	Congo-Brazzaville	132,047
26	Ethiopia	426,373	63	Finland	130,559
27	Bolivia	424,165	64	Vietnam	128,066
28	Mauritania	397,956	65	Malaysia	127,320
29	Egypt	386,662	66	Norway	125,050
30	Tanzania	364,900	67	Cote d'Ivoire	124,504
31	Nigeria	356,669	68	Poland	120,728
32	Venezuela	352,145	69	Italy	116,342
33	Pakistan	339,732	70	Philippines	115,831
34	Namibia	317,818	71	Ecuador	109,484
35	Mozambique	309,496	72	Burkina Faso	105,869
36	Turkey	302,541	73	New Zealand	104,454
37	Chile	291,930	74	Gabon	103,347

APPENDIX C
Country Rank by Area

Rank	Country	Area (miles²)	Rank	Country	Area (miles²)
75	Guinea	94,926	112	Austria	32,378
76	United Kingdom	93,788	113	United Arab Emirates	32,278
77	Uganda	93,065	114	Czech Republic	30,450
78	Ghana	92,098	115	Panama	29,157
79	Romania	91,699	116	Sierra Leone	27,699
80	Laos	91,429	117	Ireland	27,133
81	Guyana	83,000	118	Georgia	26,911
82	Oman	82,030	119	Sri Lanka	25,332
83	Belarus	80,155	120	Lithuania	25,213
84	Kyrgyzstan	77,182	121	Latvia	24,942
85	Senegal	75,951	122	Togo	21,925
86	Syria	71,498	123	Croatia	21,829
87	Cambodia	69,898	124	Bosnia & Herzegovina	19,767
88	Uruguay	67,574	125	Costa Rica	19,730
89	Tunisia	63,170	126	Slovakia	18,924
90	Suriname	63,037	127	Dominican Republic	18,730
91	Nepal	56,827	128	Bhutan	17,954
92	Bangladesh	55,598	129	Estonia	17,462
93	Tajikistan	55,251	130	Denmark	16,640
94	Greece	50,949	131	Netherlands	16,164
95	Nicaragua	50,054	132	Switzerland	15,943
96	North Korea	46,540	133	Guinea-Bissau	13,948
97	Malawi	45,747	134	Taiwan	13,901
98	Eritrea	45,406	135	Moldova	13,070
99	Benin	43,484	136	Belgium	11,787
100	Honduras	43,277	137	Lesotho	11,720
101	Liberia	43,000	138	Armenia	11,506
102	Bulgaria	42,855	139	Albania	11,100
103	Cuba	42,804	140	Solomon Islands	10,954
104	Guatemala	42,042	141	Equatorial Guinea	10,831
105	Iceland	39,769	142	Burundi	10,745
106	South Korea	38,230	143	Haiti	10,714
107	Hungary	35,919	144	Rwanda	10,169
108	Portugal	35,516	145	Macedonia	9,928
109	Jordan	34,495	146	Djibouti	8,958
110	Serbia	34,116	147	Belize	8,867
111	Azerbaijan	33,437	148	El Salvador	8,124

APPENDIX C
Country Rank by Area

Rank	Country	Area (miles^2)	Rank	Country	Area (miles^2)
149	Israel	8,019	172	Dominica	290
150	Slovenia	7,821	173	F. States of Micronesia	271
151	Fiji	7,056	174	Bahrain	267
152	Kuwait	6,880	175	Singapore	264
153	Swaziland	6,704	176	Tonga	251
154	Timor-Leste	5,743	177	St. Lucia	238
155	Bahamas	5,382	178	Palau	188
156	Montenegro	5,333	179	Andorra	181
157	Vanuatu	4,707	180	Seychelles	176
158	Qatar	4,412	181	Antigua & Barbuda	171
159	Jamaica	4,244	182	Barbados	166
160	Gambia	4,127	183	St. Vincent	150
161	Lebanon	4,016	184	Grenada	133
162	Cyprus	3,572	185	Malta	122
163	Brunei	2,226	186	Maldives	115
164	Trinidad & Tobago	1,980	187	St. Kitts & Nevis	101
165	Cape Verde	1,557	188	Marshall Islands	70
166	Samoa	1,093	189	Liechtenstein	62
167	Luxembourg	999	190	San Marino	24
168	Comoros	863	191	Tuvalu	10
169	Mauritius	788	192	Nauru	8
170	São Tomé & Principe	372	193	Monaco	.75
171	Kiribati	313	194	Vatican City	.17

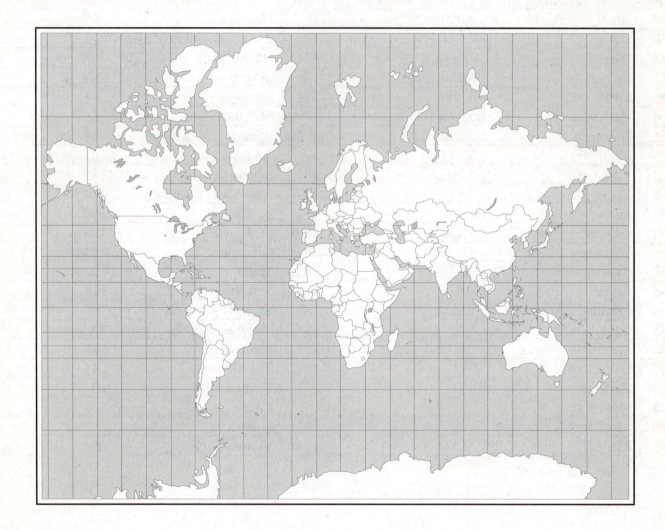

APPENDIX D
Country Rank by Population

Rank	Country	Population	Rank	Country	Population
1	China	1,321,852,000	38	Afghanistan	31,890,000
2	India	1,129,866,000	39	Uganda	30,263,000
3	United States	301,140,000	40	Nepal	28,902,000
4	Indonesia	234,694,000	41	Peru	28,675,000
5	Brazil	190,011,000	42	Uzbekistan	27,780,000
6	Pakistan	164,742,000	43	Saudi Arabia	27,601,000
7	Bangladesh	150,448,000	44	Iraq	27,500,000
8	Russia	141,378,000	45	Venezuela	26,024,000
9	Nigeria	135,031,000	46	Malaysia	24,821,000
10	Japan	127,433,000	47	North Korea	23,302,000
11	Mexico	108,701,000	48	Ghana	22,931,000
12	Philippines	91,077,000	49	Taiwan	22,859,000
13	Vietnam	85,262,000	50	Romania	22,276,000
14	Germany	82,401,000	51	Yemen	22,231,000
15	Egypt	80,335,000	52	Sri Lanka	20,926,000
16	Ethiopia	76,512,000	53	Mozambique	20,906,000
17	Turkey	71,159,000	54	Australia	20,434,000
18	Congo-Kinshasa	65,752,000	55	Madagascar	19,449,000
19	Iran	65,398,000	56	Syria	19,315,000
20	Thailand	65,068,000	57	Cameroon	18,060,000
21	France	60,876,000	58	Cote d'Ivoire	18,013,000
22	United Kingdom	60,776,000	59	Netherlands	16,571,000
23	Italy	58,148,000	60	Chile	16,285,000
24	South Korea	49,045,000	61	Kazakhstan	15,285,000
25	Burma	47,374,000	62	Burkina Faso	14,326,000
26	Ukraine	46,300,000	63	Cambodia	13,996,000
27	Colombia	44,380,000	64	Ecuador	13,756,000
28	South Africa	43,998,000	65	Malawi	13,603,000
29	Spain	40,448,000	66	Niger	12,895,000
30	Argentina	40,302,000	67	Guatemala	12,728,000
31	Tanzania	39,384,000	68	Senegal	12,522,000
32	Sudan	39,379,000	69	Zimbabwe	12,311,000
33	Poland	38,518,000	70	Angola	12,264,000
34	Kenya	36,914,000	71	Mali	11,995,000
35	Morocco	33,757,000	72	Zambia	11,477,000
36	Canada	33,390,000	73	Cuba	11,394,000
37	Algeria	33,333,000	74	Greece	10,706,000

APPENDIX D
Country Rank by Population

Rank	Country	Population	Rank	Country	Population
75	Portugal	10,643,000	112	Turkmenistan	5,097,000
76	Belgium	10,392,000	113	Eritrea	4,907,000
77	Tunisia	10,276,000	114	Georgia	4,646,000
78	Czech Republic	10,229,000	115	Norway	4,628,000
79	Serbia	10,150,000	116	Singapore	4,553,000
80	Hungary	9,956,000	117	Bosnia & Herzegovina	4,552,000
81	Guinea	9,948,000	118	Croatia	4,493,000
82	Rwanda	9,908,000	119	United Arab Emirates	4,444,000
83	Chad	9,886,000	120	Central African Repub.	4,369,000
84	Belarus	9,725,000	121	Moldova	4,320,000
85	Dominican Republic	9,366,000	122	Costa Rica	4,134,000
86	Bolivia	9,119,000	123	New Zealand	4,116,000
87	Somalia	9,119,000	124	Ireland	4,109,000
88	Sweden	9,031,000	125	Lebanon	3,926,000
89	Haiti	8,706,000	126	Congo-Brazzaville	3,801,000
90	Burundi	8,391,000	127	Albania	3,601,000
91	Austria	8,200,000	128	Lithuania	3,575,000
92	Azerbaijan	8,120,000	129	Uruguay	3,461,000
93	Benin	8,078,000	130	Mauritania	3,270,000
94	Switzerland	7,555,000	131	Panama	3,242,000
95	Honduras	7,484,000	132	Oman	3,205,000
96	Bulgaria	7,323,000	133	Liberia	3,196,000
97	Tajikistan	7,077,000	134	Armenia	2,972,000
98	El Salvador	6,948,000	135	Mongolia	2,952,000
99	Paraguay	6,669,000	136	Jamaica	2,780,000
100	Laos	6,522,000	137	Kuwait	2,506,000
101	Israel	6,427,000	138	Bhutan	2,328,000
102	Sierra Leone	6,145,000	139	Latvia	2,260,000
103	Jordan	6,053,000	140	Lesotho	2,125,000
104	Libya	6,037,000	141	Macedonia	2,056,000
105	Papua New Guinea	5,796,000	142	Namibia	2,055,000
106	Togo	5,702,000	143	Slovenia	2,009,000
107	Nicaragua	5,675,000	144	Botswana	1,816,000
108	Denmark	5,468,000	145	Gambia	1,688,000
109	Slovakia	5,448,000	146	Guinea-Bissau	1,473,000
110	Kyrgyzstan	5,284,000	147	Gabon	1,455,000
111	Finland	5,238,000	148	Estonia	1,316,000

Places of the World

APPENDIX D
Country Rank by Population

Rank	Country	Population	Rank	Country	Population
149	Mauritius	1,251,000	172	Barbados	281,000
150	Swaziland	1,133,000	173	Samoa	214,000
151	Timor-Leste	1,085,000	174	Vanuatu	212,000
152	Trinidad & Tobago	1,057,000	175	São Tomé & Principe	200,000
153	Fiji	919,000	176	St. Lucia	171,000
154	Qatar	907,000	177	St. Vincent	118,000
155	Cyprus	788,000	178	Tonga	117,000
156	Guyana	769,000	179	F. States of Micronesia	108,000
157	Comoros	711,000	180	Kiribati	108,000
158	Bahrain	709,000	181	Grenada	90,000
159	Montenegro	685,000	182	Seychelles	82,000
160	Solomon Islands	567,000	183	Dominica	72,000
161	Equatorial Guinea	551,000	184	Andorra	72,000
162	Djibouti	496,000	185	Antigua & Barbuda	69,000
163	Luxembourg	480,000	186	Marshall Islands	62,000
164	Suriname	471,000	187	St. Kitts & Nevis	39,000
165	Cape Verde	424,000	188	Liechtenstein	34,000
166	Malta	402,000	189	Monaco	33,000
167	Brunei	375,000	190	San Marino	30,000
168	Maldives	369,000	191	Palau	21,000
169	Bahamas	306,000	192	Nauru	14,000
170	Iceland	302,000	193	Tuvalu	12,000
171	Belize	294,000	194	Vatican City	820

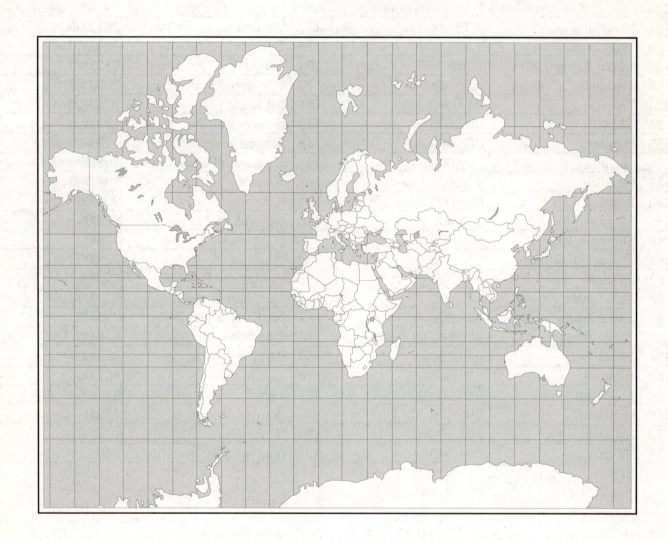

APPENDIX E
Country Rank by per capita GDP

Rank	Country	per capita GDP	Rank	Country	per capita GDP
1	Luxembourg	$71,400	38	Cyprus	$23,000
2	Equatorial Guinea	$50,200	39	Czech Republic	$22,000
3	United Arab Emirates	$49,700	40	Bahamas	$21,600
4	Norway	$46,300	41	Malta	$21,300
5	Ireland	$44,500	42	Estonia	$20,300
6	United States	$43,800	43	Portugal	$19,800
7	Andorra	$38,800	44	Trinidad & Tobago	$19,800
8	Iceland	$38,000	45	Barbados	$18,400
9	Denmark	$37,100	46	Slovakia	$18,200
10	Canada	$35,700	47	Hungary	$17,500
11	Austria	$34,700	48	Latvia	$16,000
12	San Marino	$34,100	49	Lithuania	$15,300
13	Switzerland	$34,000	50	Argentina	$15,200
14	Finland	$33,500	51	Oman	$14,400
15	Australia	$33,300	52	Poland	$14,400
16	Japan	$33,100	53	Saudi Arabia	$13,800
17	Belgium	$33,000	54	Mauritius	$13,700
18	Sweden	$32,200	55	Croatia	$13,400
19	Netherlands	$32,100	56	South Africa	$13,300
20	Germany	$31,900	57	Malaysia	$12,800
21	United Kingdom	$31,800	58	Chile	$12,600
22	Singapore	$31,400	59	Costa Rica	$12,500
23	France	$31,200	60	Libya	$12,300
24	Italy	$30,200	61	Russia	$12,200
25	Monaco	$30,000	62	Antigua & Barbuda	$10,900
26	Qatar	$29,800	63	Botswana	$10,900
27	Taiwan	$29,600	64	Uruguay	$10,900
28	Spain	$27,400	65	Bulgaria	$10,700
29	Israel	$26,800	66	Mexico	$10,700
30	New Zealand	$26,200	67	Kazakhstan	$9,400
31	Bahrain	$25,600	68	Thailand	$9,200
32	Brunei	$25,600	69	Turkey	$9,100
33	Liechtenstein	$25,000	70	Romania	$9,100
34	South Korea	$24,500	71	Tunisia	$8,900
35	Greece	$24,000	72	Brazil	$8,800
36	Slovenia	$23,400	73	Iran	$8,700
37	Kuwait	$23,100	74	Colombia	$8,600

APPENDIX E
Country Rank by per capita GDP

Rank	Country	per capita GDP	Rank	Country	per capita GDP
75	Turkmenistan	$8,500	112	Ecuador	$4,500
76	Belize	$8,400	113	Serbia	$4,400
77	Dominican Republic	$8,400	114	Egypt	$4,200
78	Macedonia	$8,300	115	Cuba	$4,100
79	Panama	$8,200	116	Syria	$4,100
80	St. Kitts & Nevis	$8,200	117	Georgia	$3,900
81	Belarus	$8,100	118	Grenada	$3,900
82	China	$7,800	119	Maldives	$3,900
83	Ukraine	$7,800	120	Indonesia	$3,900
84	Seychelles	$7,800	121	Dominica	$3,800
85	Algeria	$7,600	122	Montenegro	$3,800
86	Palau	$7,600	123	India	$3,800
87	Azerbaijan	$7,500	124	St. Vincent & Grenadines	$3,600
88	Namibia	$7,500	125	Bolivia	$3,100
89	Venezuela	$7,200	126	Vietnam	$3,100
90	Gabon	$7,100	127	Honduras	$3,100
91	Suriname	$7,100	128	Nicaragua	$3,100
92	Peru	$6,600	129	Vanuatu	$2,900
93	Fiji	$6,200	130	Marshall Islands	$2,900
94	Cape Verde	$6,000	131	Cambodia	$2,800
95	Lebanon	$5,900	132	Kiribati	$2,800
96	Albania	$5,700	133	Ghana	$2,700
97	Armenia	$5,700	134	Papua New Guinea	$2,700
98	Bosnia & Herzegovina	$5,600	135	Lesotho	$2,700
99	Swaziland	$5,300	136	Mauritania	$2,600
100	Jordan	$5,100	137	Pakistan	$2,600
101	Guatemala	$5,000	138	Cameroon	$2,500
102	Philippines	$5,000	139	Sudan	$2,400
103	Nauru	$5,000	140	Bangladesh	$2,300
104	El Salvador	$4,900	141	Fed. States of Micronesia	$2,300
105	Guyana	$4,900	142	Laos	$2,200
106	Paraguay	$4,800	143	Tonga	$2,200
107	St. Lucia	$4,800	144	Guinea	$2,100
108	Sri Lanka	$4,700	145	Mongolia	$2,100
109	Jamaica	$4,700	146	Samoa	$2,100
110	Morocco	$4,600	147	Zimbabwe	$2,100
111	Angola	$4,500	148	Kyrgyzstan	$2,100

APPENDIX E
Country Rank by per capita GDP

Rank	Country	per capita GDP	Rank	Country	per capita GDP
149	Gambia	$2,000	172	Kenya	$1,200
150	Uzbekistan	$2,000	173	São Tomé & Principe	$1,200
151	Moldova	$2,000	174	Benin	$1,100
152	Iraq	$1,900	175	Djibouti	$1,100
153	Uganda	$1,900	176	Ethiopia	$1,100
154	Burma	$1,800	177	Zambia	$1,000
155	Haiti	$1,800	178	Yemen	$1,000
156	North Korea	$1,800	179	Niger	$1,000
157	Senegal	$1,800	180	Eritrea	$1,000
158	Togo	$1,700	181	Liberia	$900
159	Cote d'Ivoire	$1,600	182	Madagascar	$900
160	Tuvalu	$1,600	183	Guinea-Bissau	$900
161	Rwanda	$1,600	184	Sierra Leone	$900
162	Chad	$1,500	185	Afghanistan	$800
163	Nepal	$1,500	186	Tanzania	$800
164	Nigeria	$1,500	187	Timor-Leste	$800
165	Mozambique	$1,500	188	Burundi	$700
166	Bhutan	$1,400	189	Congo-Kinshasa	$700
167	Burkina Faso	$1,400	190	Solomon Islands	$600
168	Congo-Brazzaville	$1,400	191	Comoros	$600
169	Mali	$1,300	192	Somalia	$600
170	Tajikistan	$1,300	193	Malawi	$600
171	Central African Republic	$1,200			

Note: This list of the world's *per capita GDP's* totals 193 rather than 194 because Vatican City is omitted (although The Holy See is a fully sovereign political entity, its ambiguous designation as a "country" means that comparable economic data are not available).

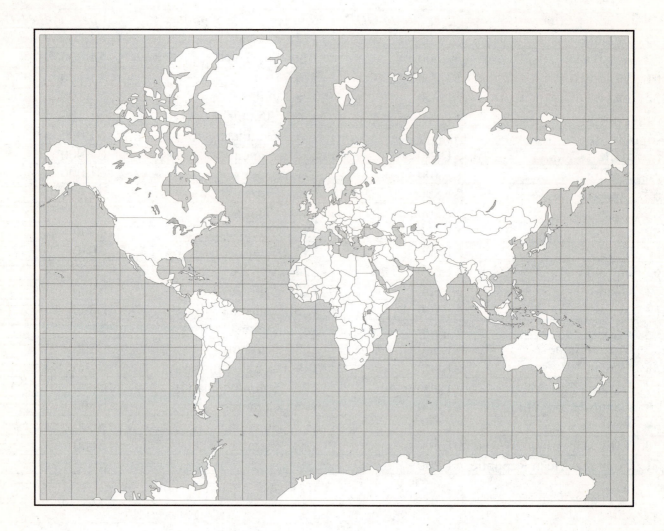

Places of the World

APPENDIX F
Alphabetical List of the World's Countries

Country	*Capital*	*Region*
Afghanistan	Kabul	*The Middle East & North Africa*
Albania	Tirana	*Europe*
Algeria	Algiers	*The Middle East & North Africa*
Andorra	Andorra la Vella	*Europe*
Angola	Luanda	*Sub-Saharan Africa*
Antigua & Barbuda	St. John's	*Latin America*
Argentina	Buenos Aires	*Latin America*
Armenia	Yerevan	*Russia and the Near Abroad*
Australia	Canberra	*Oceania*
Austria	Vienna	*Europe*
Azerbaijan	Baku	*Russia and the Near Abroad*
Bahamas	Nassau	*Latin America*
Bahrain	Manama	*The Middle East & North Africa*
Bangladesh	Dhaka	*Monsoon Asia*
Barbados	Bridgetown	*Latin America*
Belarus	Minsk	*Russia and the Near Abroad*
Belgium	Brussels	*Europe*
Belize	Belmopan	*Latin America*
Benin	Porto-Novo	*Sub-Saharan Africa*
Bhutan	Thimphu	*Monsoon Asia*
Bolivia	Sucre & La Paz	*Latin America*
Bosnia-Herzegovina	Sarajevo	*Europe*
Botswana	Gaborone	*Sub-Saharan Africa*
Brazil	Brasilia	*Latin America*
Brunei	Bandar Seri Begawan	*Monsoon Asia*
Bulgaria	Sofia	*Europe*
Burkina Faso	Ouagadougou	*Sub-Saharan Africa*
Burma	Rangoon	*Monsoon Asia*
Burundi	Bujumbura	*Sub-Saharan Africa*
Cambodia	Phnom Penh	*Monsoon Asia*
Cameroon	Yaoundé	*Sub-Saharan Africa*
Canada	Ottawa	*North America (U.S. & Canada)*
Cape Verde	Praia	*Sub-Saharan Africa*

Country	Capital	Region
Central African Republic	Bangui	*Sub-Saharan Africa*
Chad	N'Djamena	*Sub-Saharan Africa*
Chile	Santiago	*Latin America*
China	Beijing	*Monsoon Asia*
Colombia	Bogota	*Latin America*
Comoros	Moroni	*Sub-Saharan Africa*
Congo	Kinshasa	*Sub-Saharan Africa*
Congo Republic	Brazzaville	*Sub-Saharan Africa*
Costa Rica	San José	*Latin America*
Côte d'Ivoire	Yamoussoukro & Abidjan	*Sub-Saharan Africa*
Croatia	Zagreb	*Europe*
Cuba	Havana	*Latin America*
Cyprus	Nicosia	*Europe*
Czech Republic	Prague	*Europe*
Denmark	Copenhagen	*Europe*
Djibouti	Djibouti	*Sub-Saharan Africa*
Dominica	Rosseau	*Latin America*
Dominican Republic	Santo Domingo	*Latin America*
Ecuador	Quito	*Latin America*
Egypt	Cairo	*The Middle East & North Africa*
El Salvador	San Salvador	*Latin America*
Equatorial Guinea	Malabo	*Sub-Saharan Africa*
Eritrea	Asmara	*Sub-Saharan Africa*
Estonia	Tallinn	*Europe*
Ethiopia	Addis Ababa	*Sub-Saharan Africa*
Fiji	Suva	*Oceania*
Finland	Helsinki	*Europe*
France	Paris	*Europe*
Gabon	Libreville	*Sub-Saharan Africa*
Gambia	Banjul	*Sub-Saharan Africa*
Georgia	T'bilisi	*Russia and the Near Abroad*
Germany	Berlin	*Europe*
Ghana	Accra	*Sub-Saharan Africa*
Greece	Athens	*Europe*
Grenada	St. George's	*Latin America*
Guatemala	Guatemala City	*Latin America*

Country	Capital	Region
Guinea	Conakry	*Sub-Saharan Africa*
Guinea-Bissau	Bissau	*Sub-Saharan Africa*
Guyana	Georgetown	*Latin America*
Haiti	Port-au-Prince	*Latin America*
Honduras	Tegucigalpa	*Latin America*
Hungary	Budapest	*Europe*
Iceland	Reykjavik	*Europe*
India	New Delhi	*Monsoon Asia*
Indonesia	Jakarta	*Monsoon Asia*
Iran	Tehran	*The Middle East & North Africa*
Iraq	Baghdad	*The Middle East & North Africa*
Ireland	Dublin	*Europe*
Israel	Jerusalem	*The Middle East & North Africa*
Italy	Rome	*Europe*
Jamaica	Kingston	*Latin America*
Japan	Tokyo	*Monsoon Asia*
Jordan	Amman	*The Middle East & North Africa*
Kazakhstan	Astana	*Russia and the Near Abroad*
Kenya	Nairobi	*Sub-Saharan Africa*
Kiribati	Tarawa	*Oceania*
Kuwait	Kuwait City	*The Middle East & North Africa*
Kyrgyzstan	Bishkek	*Russia and the Near Abroad*
Laos	Vientiane	*Monsoon Asia*
Latvia	Riga	*Europe*
Lebanon	Beirut	*The Middle East & North Africa*
Lesotho	Maseru	*Sub-Saharan Africa*
Liberia	Monrovia	*Sub-Saharan Africa*
Libya	Tripoli	*The Middle East & North Africa*
Liechtenstein	Vaduz	*Europe*
Lithuania	Vilnius	*Europe*
Luxembourg	Luxembourg-ville	*Europe*
Macedonia	Skopje	*Europe*
Madagascar	Antananarivo	*Sub-Saharan Africa*
Malawi	Lilongwe	*Sub-Saharan Africa*

Country	Capital	Region
Malaysia	Kuala Lumpur	*Monsoon Asia*
Maldives	Male	*Monsoon Asia*
Mali	Bamako	*Sub-Saharan Africa*
Malta	Valletta	*Europe*
Marshall Islands	Majuro	*Oceania*
Mauritania	Nouakchott	*Sub-Saharan Africa*
Mauritius	Port Louis	*Sub-Saharan Africa*
Mexico	Mexico City	*Latin America*
Micronesia	Palikir	*Oceania*
Moldova	Chisinau	*Russia and the Near Abroad*
Monaco	Monaco-ville	*Europe*
Mongolia	Ulaanbaatar	*Monsoon Asia*
Montenegro	Podgorica	*Europe*
Morocco	Rabat	*The Middle East & North Africa*
Mozambique	Maputo	*Sub-Saharan Africa*
Namibia	Windhoek	*Sub-Saharan Africa*
Nauru	Yaren District	*Oceania*
Nepal	Kathmandu	*Monsoon Asia*
Netherlands	Amsterdam & The Hague	*Europe*
New Zealand	Wellington	*Oceania*
Nicaragua	Managua	*Latin America*
Niger	Niamey	*Sub-Saharan Africa*
Nigeria	Abuja	*Sub-Saharan Africa*
North Korea	Pyongyang	*Monsoon Asia*
Norway	Oslo	*Europe*
Oman	Muscat	*The Middle East & North Africa*
Pakistan	Islamabad	*Monsoon Asia*
Palau	Melekeok	*Oceania*
Panama	Panama City	*Latin America*
Papua New Guinea	Port Moresby	*Oceania*
Paraguay	Asunción	*Latin America*
Peru	Lima	*Latin America*
Philippines	Manila	*Monsoon Asia*
Poland	Warsaw	*Europe*
Portugal	Lisbon	*Europe*
Qatar	Doha	*The Middle East & North Africa*

Country	Capital	Region
Romania	Bucharest	*Europe*
Russia	Moscow	*Russia and the Near Abroad*
Rwanda	Kigali	*Sub-Saharan Africa*
St. Kitts and Nevis	Basseterre	*Latin America*
St. Lucia	Castries	*Latin America*
St. Vincent/Grenadines	Kingstown	*Latin America*
Samoa	Apia	*Oceania*
San Marino	San Marino	*Europe*
São Tomé & Príncipe	São Tomé	*Sub-Saharan Africa*
Saudi Arabia	Riyadh	*The Middle East & North Africa*
Senegal	Dakar	*Sub-Saharan Africa*
Serbia	Belgrade	*Europe*
Seychelles	Victoria	*Sub-Saharan Africa*
Sierra Leone	Freetown	*Sub-Saharan Africa*
Singapore	Singapore	*Monsoon Asia*
Slovakia	Bratislava	*Europe*
Slovenia	Ljubljana	*Europe*
Solomon Islands	Honiara	*Oceania*
Somalia	Mogadishu	*Sub-Saharan Africa*
South Africa	Pretoria	*Sub-Saharan Africa*
South Korea	Seoul	*Monsoon Asia*
Spain	Madrid	*Europe*
Sri Lanka	Colombo	*Monsoon Asia*
Sudan	Khartoum	*The Middle East & North Africa*
Suriname	Paramaribo	*Latin America*
Swaziland	Mbabane	*Sub-Saharan Africa*
Sweden	Stockholm	*Europe*
Switzerland	Bern	*Europe*
Syria	Damascus	*The Middle East & North Africa*
Taiwan	Taipei	*Monsoon Asia*
Tajikistan	Dushanbe	*Russia and the Near Abroad*
Tanzania	Dar es Salaam	*Sub-Saharan Africa*
Thailand	Bangkok	*Monsoon Asia*
Timor-Leste	Dili	*Monsoon Asia*
Togo	Lomé	*Sub-Saharan Africa*
Tonga	Nuku'alofa	*Oceania*
Trinidad & Tobago	Port of Spain	*Latin America*
Tunisia	Tunis	*The Middle East & North Africa*

Country	Capital	Region
Turkey	Ankara	*The Middle East & North Africa*
Turkmenistan	Ashgabat	*Russia and the Near Abroad*
Tuvalu	Funafuti	*Oceania*
Uganda	Kampala	*Sub-Saharan Africa*
Ukraine	Kiev	*Russia and the Near Abroad*
United Arab Emirates	Abu Dhabi	*The Middle East & North Africa*
United Kingdom	London	*Europe*
United States of America	Washington, D.C.	*North America (U.S. & Canada)*
Uruguay	Montevideo	*Latin America*
Uzbekistan	Tashkent	*Russia and the Near Abroad*
Vanuatu	Port Vila	*Oceania*
Vatican City	Vatican City	*Europe*
Venezuela	Caracas	*Latin America*
Vietnam	Hanoi	*Monsoon Asia*
Yemen	Sanaa	*The Middle East & North Africa*
Zambia	Lusaka	*Sub-Saharan Africa*
Zimbabwe	Harare	*Sub-Saharan Africa*

APPENDIX G:
Updates & Corrections

My experience from writing the first five editions of this book is that careful proofreading will uncover most typographical and factual errors, but the rest will wait until the moment the book has been printed and bound before immediately revealing themselves. In addition, of course, the world is constantly changing, so that much of the data presented in these pages are ephemeral by their nature. The result of these two unfortunate phenomena is that this book will inevitably be in need of updates and corrections, and that need will almost certainly arise sooner rather than later.

As I become aware of important changes that are required for this sixth edition, I will publish them on my faculty webpage, where they will be available free of charge as a downloadable file. To access the latest updates and corrections, please visit http://faculty.ircc.edu/faculty/jlett. From the homepage, choose the link to "Publications," and then select the link for *Places of the World Updates & Corrections* under the heading "Books".

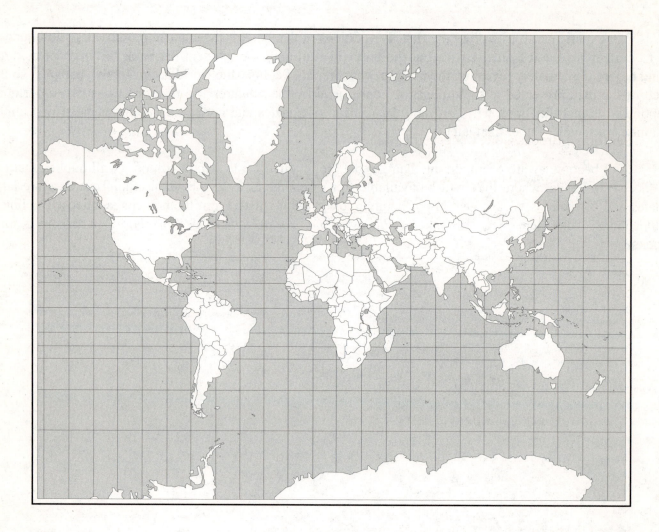

Places of the World

BIBLIOGRAPHIC AND ONLINE RESOURCES

The data in this book were compiled from a wide variety of sources; some of the more important ones are listed below.

Commonwealth of Nations. http://www.thecommonwealth.org/

Essentials of World Regional Geography by Joseph J. Hobbs. 2009. 6th Edition. Pacific Grove, California: Brooks/Cole, Cengage Learning.

Europa: Gateway to the European Union. http://europa.eu./index_en.htm

Goode's World Atlas. 21st edition. 2006. Rand McNally.

League of Arab States. http://www.arableagueonline.org/

Merriam-Webster's Geographical Dictionary. 3rd Edition. 1998. Springfield, Massachusetts: Merriam Webster.

North Atlantic Treaty Organization. http://www.nato.int/

Organization of American States. http://www.oas.org/

Organization of Petroleum Exporting Countries. http://www.opec.org/

Rand McNally Atlas of World Geography. 2005. Rand McNally.

United Nations. http://www.un.org/english/

United States Board on Geographic Names. http://geonames.usgs.gov/

United States Department of State. http://www.state.gov/countries/

The World Factbook 2008. http://www.cia.gov/library/publications/the-world-factbook/

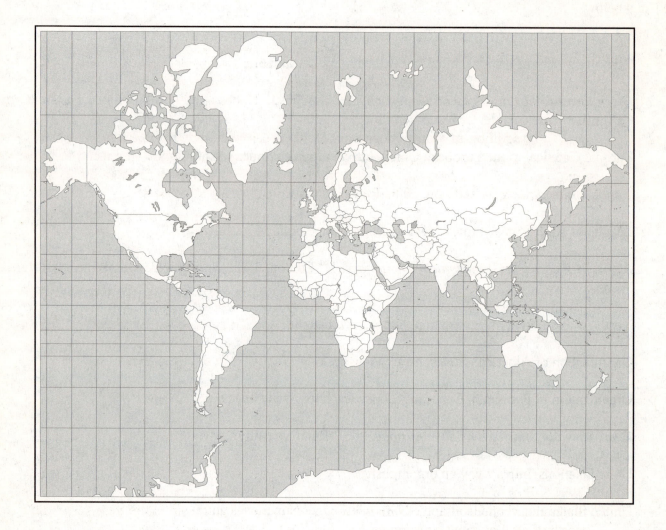

ABOUT THE AUTHOR

James Lett is Honors Professor of Anthropology and Geography at Indian River College in Ft. Pierce, Florida, where he served as Chair of the Department of Social Sciences from 1993 to 1998. He was educated at the College of William and Mary (B.A., 1977) and the University of Florida (Ph.D., 1983), and he has taught at several colleges and universities, including the University of Florida, Florida Atlantic University, the Florida Institute of Technology, and the University of Utah.

In addition to world geography, Dr. Lett has wide-ranging interests in biological anthropology (especially hominid evolution and forensic anthropology), archaeology (especially world prehistory), ethnology (especially human universals and evolutionary psychology), and linguistics (especially the intersection between language and the evolution of the human mind). He is the author of two books on anthropological theory, *Science, Reason, and Anthropology—The Principles of Rational Inquiry* (Rowman and Littlefield, 1997) and *The Human Enterprise—A Critical Introduction to Anthropological Theory* (Westview Press, 1987), and co-author of several others, including *Studying Societies and Cultures* (2007), *Encyclopedia of Religious Rites, Rituals, and Festivals* (2004), *Classics of Practicing Anthropology* (2000), *Encounters with the Paranormal—Science, Knowledge, and Belief* (1998), *Anthropology of Religion—A Handbook* (1997), *Encyclopedia of Cultural Anthropology* (1996), *Media Anthropology* (1994), *The Hundredth Monkey and Other Paradigms of the Paranormal* (1991), *Emics and Etics—The Insider/Outsider Debate* (1990), and *Hosts and Guests—The Anthropology of Tourism* (1989).

Since 1998, James Lett has been the host and producer of *Excursions in Geography*, a daily one-minute radio segment broadcast on NPR-Affiliate WQCS-FM in Ft. Pierce, Florida. In 2007, Pioneer Press published a fully illustrated version of *Excursions in Geography* which features an entire year's worth of Dr. Lett's radio programs.

Professor Lett has lived abroad (in Europe, Asia, and the Caribbean) for a total of nearly five years, and he continues to travel extensively. He currently makes his home on the southeast coast of Florida, where his recreational activities include boating, golf, and tennis.

Additional information about James Lett is available on his website at http://faculty.ircc.edu/faculty/jlett